SPACE STATIONS

SPACE

STATIONS

Base Camps

to the Stars

ROGER D. LAUNIUS

Smithsonian Books

Washington and London

© 2003 Smithsonian Institution
All rights reserved

Copy Editor: Anne Collier Rehill
Production Editor: Ruth G. Thomson
Designer: Brian Barth

Library of Congress Cataloging-in-Publication Data
Launius, Roger D.
    Space stations : base camps to the stars / Roger D. Launius
          p.   cm.
    Includes bibliographical references (p.) and index.
    ISBN 1-58834-120-8 (alk. paper)
     1. International Space Station. 2. Space stations. I. Title.
    TL 797.L38   2002
    629.44r2—dc21   2002036508

British Library Cataloging-in-Publication Data available

Manufactured in China
10 09 08 07 06 05 04 03   5 4 3 2 1

The paper used in this publication meets the minimum require-
ments of the American National Standard for Information
Sciences—Permanence of Paper for Printed Library Materials
ANSI Z39.48-1984.

For permission to reproduce illustrations appearing in this book,
please correspond directly with the owners of the works, as listed
in the individual captions. Smithsonian Books does not retain
reproduction rights for these illustrations individually or maintain
a file of addresses for photo sources.

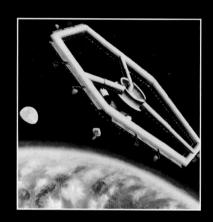
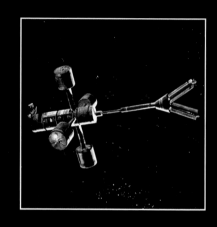

CONTENTS

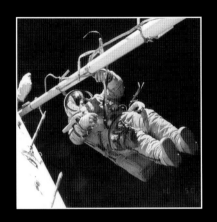

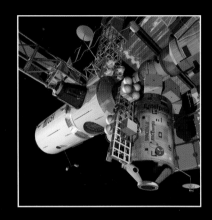

One of the uniquely powerful images of spaceflight throughout the twentieth century has been that of a space station floating above Earth and serving as a way station to the universe beyond. Sometimes this has been envisioned in art, literature, or film as an extravagant transition point—as in science fiction films from *2001: A Space Odyssey* to *Star Trek* to *Mission to Mars*—and at other times it has been more realistically depicted as a spare research outpost. But no matter how it is illustrated, the space station has been a central goal of the spacefaring community of the United States. The belief has existed from the beginning of space exploration that once a rocket overcomes Earth's gravity and reaches orbit, travelers are "halfway to anywhere" they might want to go. At such a mythical halfway point, space stations could serve as transit points between vehicles traveling back and forth between Earth and the Moon and planets.

In this context, a space station was seen as a "base camp" for further exploration. It has been touted by early visionaries such as Konstantin E. Tsiolkovskiy, Hermann Oberth, and Hermann Noordung, as well as by recent advocates of humanity's movement into space. From the dreams of those early pioneers through the studies of the 1960s, Skylab, President Reagan's approval to proceed with a station, the Freedom years, Mir, and the ultimate completion of the project, this commitment has been constant.

This book seeks to explore the power of the space station as an icon of American spacefaring culture throughout the twentieth century. Starting with the first known mention of a station, in a short story in *Atlantic Monthly* magazine in 1869, the book follows the project through conceptualization to Wernher von Braun's expansive vision in the 1950s, through today and beyond. From its earliest beginnings, the space station has been considered critical for further exploration of the Moon and planets.

In part because of the persistent vision of human destiny to explore the solar system and the central role of a space station in facilitating this goal, studies of station configurations were an important part of the National Aeronautics and Space Administration's planning as early as the 1960s. NASA scientists and engineers pressed for these studies because a space station met the needs of the agency for an orbital laboratory, observatory, industrial plant, launching platform, and drydock. But the station was forced to the bottom of the priority heap in 1961, with the Kennedy decision to land an American on the Moon by the end of the decade. With that mandate, there was no time to develop a station in spite of the fact that virtually everyone in NASA recognized its use for exploration beyond Earth orbit. It surfaced as the foremost NASA program even while Apollo became a reality in the latter 1960s, but took several unusual turns before emerging as the principal project of the agency at the end of the twentieth century.

The Soviet/Russian succession of Salyut stations and Mir complemented the work of the United States to develop what is now called the International Space Station (ISS). With the first element launches of this station in 1998 and the occupation by the first crew in 2000, the spacefaring nations of the world intend that no future generation will know a time when there is not some human presence in space. This book elucidates the history of this vision of space exploration and links its realization in the ISS to the longstanding cultural conception of what it means to be a spacefaring people.

ACKNOWLEDGMENTS

Any book incurs numerous debts over the course of its production. This is especially true of a breathless survey such as this, which of necessity relies almost exclusively on the primary research of others. I wish to acknowledge the support and encouragement of a large number of people associated with aerospace studies, and I want to thank many individuals who materially contributed to the completion of this project. Of course, we would never have taken on this project except for the encouragement and ideas provided by Mark Gatlin and the other fine people at Smithsonian Books who worked on it, especially Anne Collier Rehill, Ruth Thomson, and Matt Litts.

In addition, several individuals read all or part of the manuscript or otherwise offered suggestions that helped me more than they will ever know. My thanks are extended to the staff of the NASA History Office: Stephen J. Garber, who offered constructive suggestions; Jane Odom, who helped track down materials and correct inconsistencies; M. Louise Alstork, who provided excellent editorial advice; Nadine Andreassen, who offered invaluable assistance; Colin Fries, who with cheerful determination helped find photographs and documents; and John Hargenrader, who proved priceless in digging out obscure details about space station designs. In addition to these individuals, the following aided me in a variety of ways: Jeff Bingham, Renee Bouchard, Michael L. Ciancone, Tom D. Crouch, Virginia P. Dawson, Frederick C. Durant III, Donald C. Elder, Richard Foust, Lori B. Garver, Michael H. Gorn, Adam L. Gruen, Richard P. Hallion, T. A. Heppenheimer, Frank Hoban, Nancy M. House, Diana Hoyt, Dennis R. Jenkins, W. D. Kay, Sylvia K. Kraemer, Alan M. Ladwig, W. Henry Lambright, Elaine Liston, John M. Logsdon, John L. Loos, Howard E. McCurdy, John E. Naugle, Allan A. Needell, Arnauld S. Nicogossian, Frederick I. Ordway III, Andrew Pedrick, Gwen Pittman, Tony Springer, Rick W. Sturdevant, Glen E. Swanson, Bert Ulrich, and Joni Wilson.

SPACE STATIONS

CHAPTER I

Origins

C uriosity about the universe and other planets has been a constant in the history of humankind. Before the twentieth century, there was little opportunity to explore the universe except in fiction and through astronomical observations. Such early probes led to the compilation of a body of knowledge that inspired and in some respects informed the efforts of scientists and engineers who in the early 1900s began to think about applying rocket technology to the challenge of spaceflight. These individuals were essentially the pioneers, translating centuries of dreams into a reality that matched in some measure the expectations of the public that watched and the governments that supported their efforts.[1]

From virtually the moment when human beings realized that flying in space was a possibility, they also understood that the building of an Earth-orbiting space station was necessary. It would serve as a location for performing all manner of scientific research, both on the Earth and in the space environment. It would facilitate conducting studies that would eventually make it possible for humans to survive in the new environment. Most important, perhaps, it would serve as the jumping-off point to the Moon and planets. Those were destinations to which many believed humanity was preordained to go, to explore, and eventually to colonize.

Space exploration enthusiasts have always believed a permanently occupied station to be a necessary outpost in the new frontier. The more technically minded have also recognized that once humans achieved Earth orbit above 200 miles in altitude (the presumed location of any space station), the vast majority of the atmosphere and the steep gravity well would be conquered. With the establishment of such an outpost, people would be about halfway to anywhere they might want to go.

The dream of a space station crystallized in the first part of the twentieth century. Between the 1920s and the early 1950s, it evolved into a concrete set of rationales, concepts, and finally proposals. In every case, it was aided by the emerging reality of spaceflight as a legitimate technological endeavor that humanity could pursue.[2] As an example of how these perspectives converged, in December 1949 Gallup pollsters found that only 15 percent of Americans believed humans could reach the Moon within fifty years. A whopping 70 percent believed it would not happen within that time, whereas 15 percent had no opinion. By October 1957, at the same time as the launching of the Soviet satellite Sputnik 1, only 25 percent believed it would take longer than twenty-five years for humanity to reach the Moon, whereas 41 percent believed firmly that it would happen within that time, and 34 percent were unsure. An important shift in perceptions took place during that era, and it was largely the result of well-known advances in rocket technology coupled with a public-relations campaign based on the real possibility of spaceflight.[3]

The discussion that follows explores that process of transformation from the dream to the possibility. It relates the critical role of several key individuals who had concrete ideas on space stations and their role in human flight but who at the time were viewed as cranks or worse. Their ideas found expression in a succession of efforts to create a permanent presence in Earth orbit.

EDWARD EVERETT HALE AND "THE BRICK MOON"

It became an article of faith for the spaceflight gospel that humanity had to build a massive Earth-orbital space station that would serve as the jumping-off point to the Moon and planets. Space exploration enthusiasts had always believed this, but the station as an idea emerged in popular culture before it became a concept in spaceflight doctrine.

Artist's conception of the "Brick Moon." (NASA photo, NASA Historical Reference Collection, NASA Headquarters, Washington, D.C.)

In 1869 Edward Everett Hale, a New England writer and social critic, published a short story in the *Atlantic Monthly* titled "The Brick Moon." The first known proposal for an orbital satellite around the Earth, Hale's piece described how a satellite in polar orbit could be used as a navigational aid to oceangoing vessels. Since the North Star offered an excellent sighting for determining latitude, perhaps something fixed in polar orbit around the Earth could provide a corresponding reference for longitude. "For you see that if, by good luck," one of Hale's characters explained,

there were a ring like Saturn's which stretched round the world, above Greenwich and the meridian of Greenwich . . . any one who wanted to measure his longitude or distance from Greenwich would look out of his window and see how high this ring was above his horizon. . . . So if we only had a ring like that . . . vertical to the plane of the equator, as the brass ring of an artificial globe goes, only far higher in proportion . . . we could calculate the longitude.[4]

The heroes of Hale's story—George Orcutt, Frederic Ingham, Ben Brannan, George Haliburton, and Gnat Q. Hale—set out to build such a navigational device, without success. Instead they decide—in the first recorded design change in spaceflight history—that such a polar ring is not feasible. They substitute plans for a single Earth-orbiting satellite on which sailing vessels can take a "fix" and thereby learn their position on the Earth's surface.

This plan would have resolved a longstanding problem that had vexed sailors for centuries, leading to the deaths of thousands as ships strayed off course and were lost at sea.[5] The building of this brick ball to be sent into orbit—brick because it can withstand fire—occupies much of the rest of the *Atlantic Monthly* story. The builders propose hurling the ball into orbit 4,000 miles above the Earth, using a complex system of two gigantic flywheels much like a modern pitching machine powered by massive waterfalls. After years of constant turning, the characters determine, the flywheels will store up enough energy to catapult the brick moon into orbit.

Unfortunately, an accident sends it off prematurely. Built at the upper end of a canal, from where it will slide to the flywheels for launch, the satellite heads for the wheels when its support structure fails before completion. The launch is perfect, as the heroes of the story knew would be the case, but thirty-seven individuals are inside at launch. George Orcutt is taking friends on a midnight tour of the structure when it catapults into space, and these individuals become the first astronauts. In contrast to what is now known about the vacuum of space, these people are able to establish a new civilization in Earth orbit and live on and in the brick moon. They raise food and enjoy an almost utopian existence.

For a year after launch, Ingham, Haliburton, and most of the rest of the Earth's scientific population search the sky for the brick moon, and Professor Karl Zitta of Breslau eventually finds it. As discoverer, Zitta names the new heavenly body, calling it Phoebe. Observations of the brick moon follow throughout the world, and astronomers learn that the thirty-seven people aboard are jumping up and down to create Morse code messages for Earth. The messages note that air, water, and other supplies are plentiful. The inhabitants have also planted crops ranging from wheat to banana trees and are enjoying remarkable yields, with ten harvests per year. For communication from Earth, Ingham and Haliburton decide to use huge black letters placed on snow, telling the spacefarers what is happening back home.

In 1869, few who read Hale's "Brick Moon" in the *Atlantic Monthly* viewed it as anything more than fantasy. Indeed, much of it was clearly in that category, but it foreshadowed many of the beliefs about the need for a space station and some of the required technologies. The flywheel approach for launch had been considered, although the velocity that may be attained in that manner is far less than required to reach orbit. A modern equivalent of it, however, is the MagLifter launch facility envisioned for the middle part of the twenty-first century. This will launch space vehicles horizontally along a track, using magnetic levitation to eliminate friction and linear electric motors to accelerate the vehicles to more than 1,000 mph. The MagLifter's catapult will provide launch assist to a variety of vehicles, and its hovering carriage will accelerate on a guideway using electromagnetic fields for a length of 2 to 5 miles. Near the end of the guideway, it will ascend to an angle of 55 degrees toward the east and send the vehicle into Earth orbit.[6]

Additionally, the use of satellites for navigation has solved the Earth's longitude problem, just as envisioned by Hale. Although not exactly what he had in mind in 1869, the Global Positioning System (GPS), a constellation of transponder satellites at fixed points in space that triangulate position down to meters, emerged in the 1990s as the second major commercial space application. It is now ubiquitous in air- and sea-vehicles and is becoming standard equipment for automobiles and trucks.[7] The ability to grow plants in orbit has also been proven, and, with the right mix of plants and animals, a reasonable biosphere could be maintained.[8] Finally, communications have been immeasurably enhanced because of spacecraft in Earth orbit. Although the Morse code messages of the Hale story appear quaint at the dawn of the twenty-first century, space-based communication is omnipresent. Indeed, life would be remarkably different if this capability suddenly disappeared.[9]

In some measure, Edward Everett Hale anticipated many of the concepts of a space station without envisioning its primary role as base camp to the stars. But those ideas would find expression through the work of other theorists.

PIONEERS OF SPACEFLIGHT

Virtually every theoretician of spaceflight in the first half of the twentieth century advocated a space station as a base camp to the stars. They did so for a very practical reason. The laws of physics, combined with the limited thrust of available propulsion systems then available, were insufficient to propel a spacecraft from the Earth's surface even to the Moon. One early pioneer calculated—incorrectly, as it turned out—that a launch vehicle bound directly from the Earth to the Moon would have to burn 105 tons of liquid propellant during the first second of flight just to heave itself off the ground. "Landing on the moon is beyond the borderline of what chemical fuels can do," observed spaceflight promoter Willy Ley. "The direct trip to the neighboring planets is even further beyond." Some sort of refueling station was obviously required. The space station, Ley observed, would provide the much-needed "cosmic stepping stone for spaceships which were too weak to reach another planet directly."[10]

Konstantin E. Tsiolkovskiy (1857–1935) as he appeared near the end of his life. Tsiolkovskiy's work proved critical to the development of spaceflight in the Soviet Union and enjoyed broad study in the 1950s and 1960s, as Americans sought to understand how the Soviet Union had accomplished such unexpected success in its early efforts in spaceflight. American space scientists realized upon reading Tsiolkovskiy that his theoretical efforts had provided an important basis for the development of the practical rocketry that underlay the Soviet space program. (NASA photo)

The first spaceflight pioneer to advocate a space station was a Russian schoolteacher named Konstantin E. Tsiolkovskiy. One of the most original thinkers in the history of spaceflight, Tsiolkovskiy studied the possibility of establishing a space station in Earth orbit even before 1900. He viewed it as a halfway point for space exploration, the necessary jumping-off point to the Moon and Mars. He even argued for the feasibility of building a dramatic wheeled space station that rotated slowly, to approximate gravity with centrifugal force.[11]

Born on September 17, 1857, in the village of Izhevskoye, Spassk District, Ryazan Gubernia, Tsiolkovskiy became enthralled as a boy with the possibilities of interplanetary travel, and at age fourteen started independent study using his father's library books on natural science and mathematics. He also developed a passion for invention and constructed balloons, propelled carriages, and other instruments. To further his education, his parents sent young Tsiolkovskiy to Moscow to pursue technical studies. He stayed there only three years, returning home to become a tutor in mathematics and physics. In 1878 he passed the required examinations and received a diploma to pursue work as a "people's school teacher," a teacher in essentially the Russian equivalent of an American high school. He secured a teaching position in arithmetic and geometry at the district school in Borosck, Kaluga Province, north of Moscow. He would remain in the Kaluga area for the rest of his life.

Very early Tsiolkovskiy demonstrated genius in scientific matters. In 1881, for instance, he broke new ground with an article on the fundamentals of the kinetic theory of gases. His second publication, "The Mechanics of a Living Organism," brought with it his election into the Society of Physics and Chemistry in St. Petersburg. Others, "The Problem of Flying by Means of Wings" (1890–91) and "Elementary Studies of the Airship and Its Structure" (1898), showed Tsiolkovskiy's growing fascination with flight.

Tsiolkovskiy first started writing on space in 1898, when he submitted for publication to the Russian journal *Nauchnoye Obozreniye* (Science review) a work based on years of calculations that laid out many of the principles of modern spaceflight. This article, "Investigating Space with Rocket Devices," finally appeared in 1903, opening the door to future writings on the subject. In it, Tsiolkovskiy described in depth the use of rockets for launching orbital space ships. He also established the fundamentals of orbital mechanics and proposed the then-radical use of both liquid oxygen and liquid hydrogen as fuel.[12]

There followed a series of increasingly sophisticated studies on the technical aspects of spaceflight. In the 1920s and 1930s, Tsiolkovskiy proved especially productive, publishing four major works in the earlier decade and six in the latter, elucidating the nature of bodies in orbit, developing scientific principles behind reaction vehicles, designing orbital space stations, and promoting interplanetary travel. He also furthered studies on many principles commonly used in rockets today: specific impulse to gauge engine performance, multistage boosters, and fuel mixtures such as liquid hydrogen and liquid oxygen. He covered the problems and possibilities inherent in microgravity, the promise of solar power, and space suits for extravehicular activity. In an oft-repeated quote, Tsiolkovskiy stated in 1911, "The Earth is the cradle of the mind, but we cannot remain forever in the cradle."[13]

Significantly, he never had the resources—nor perhaps the inclination—to experiment with rockets himself. Among his many studies, Tsiolkovskiy advocated the importance of an orbital space station for exploration beyond the Earth. In a 1878–79 series of three articles titled "Free Space,"

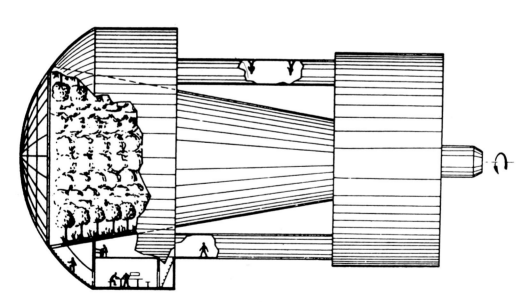

In addition to his work on rocketry, Konstantin Tsiolskovskiy's studies of a space station led to this 1903 design of a rotating space platform to create artificial gravity. The habitat located at the top of the "ice cream cone" propulsion section has an internal forest, shown in the cutaway. (NASA photo)

originally published as science fiction stories, Tsiolkovskiy proposed a space station shaped like a wheel. Inspired to adopt this design by the rings of Saturn, Tsiolkovskiy opined, "An artificial ring allows a person to move freely in all directions." This station would have gravity generated through centrifugal force but could be varied from normal Earth conditions to weightlessness, depending on the moods of the inhabitants. And it would not be limited to Earth orbit but could be placed in orbit anywhere, to facilitate studying the body that it circumnavigated.[14]

At a fundamental level, for Tsiolkovskiy the space station became the principal mode of exploration, with rockets, spacecraft, and necessary transfer vehicles decidedly secondary. The station would be the place to live and work, like a home, a factory, or a city. The rocket would provide mobility, as did a railroad, but little more. In 1912, Tsiolkovskiy predicted the creation of multiple space stations. "Piece by piece," he wrote, "colonies will be formed with materials, machinery, and structures brought from Earth. Then, an independent, albeit limited, production could gradually develop."[15]

Tsiolkovskiy visualized a space station as a small planet, a complete biosphere, but one divided into two major sections. The first would provide habitation for the crew, with space for fifty. Centrifugal force would provide artificial gravity, and there would even be a few luxuries, such as a gymnasium. A second section would provide the space needed for growing plants, and perhaps even a few animals. Tsiolkovskiy believed these tidy, urban gardens could yield all that the crew required. Waste products would be recycled for fertilizer, and urine could be purified for reuse as water. The plants would give not only food but also oxygen through photosynthesis. Energy from the Sun would enable all of this to remain active. Such a self-contained biosphere, Tsiolkovskiy believed, would make it possible for humanity to survive indefinitely in space. Without it, he reasoned, constant resupply from Earth would be necessary and would limit human exploration by tying an umbilical cord from the space station back to Earth. He also proposed harvesting asteroids for their minerals and other commodities. These raw materials could then be used to manufacture all manner of items desired on Earth. Thus the space station could be self-sufficient, both economically and practically.[16]

The station represented an essential step in practically every settlement and exploration scheme that Tsiolkovskiy offered during his long career. A station in space, he noted, would provide important knowledge about the practical details of maintaining human life away from the shelter of the Earth, to say nothing of its value as a transfer point. The knowledge gained from developing a space station would also foster the technology necessary to build self-supporting bases in more remote places, deeper in space.

Tsiolkovskiy included many of his ideas in a 1935 film produced in the Soviet Union, depicting a voyage to the Moon. Modeled after the classic 1929 German film *Frau im Mond*

(Woman on the Moon), the Soviet film organization Mosfilm decided to produce a work just as spectacular, the silent movie *Kosmicheskoye Putechestviye* (Space journey). For professional help and advice, film director V. N. Zhuravlev turned to Konstantin Tsiolkovskiy. At least from a technical standpoint, the movie was more accurate than its German counterpart, since the astronauts wore space suits on the Moon. It also was the first Soviet film that showed weightlessness in action. Thin cables, invisible to the camera, suspended the actors.[17] It represented the high-water mark of Tsiolkovskiy's public presentation of ideas, ideas that would be partially realized during the space age that began in 1957.

After the Bolshevik revolution of 1917 and the creation of the Soviet Union, Tsiolkovskiy was formally recognized for his accomplishments in the theory of spaceflight. Among other honors, in 1921 he received a lifetime pension from the state that allowed him to retire from teaching at age sixty-four. Thereafter he devoted full time to developing his spaceflight studies. He died at his home in Kaluga on September 19, 1935, two days after his seventy-eighth birthday. Tsiolkovskiy's theoretical work greatly influenced later rocketeers, both in his native land and throughout Europe. He was less well known during his lifetime in the United States, but Tsiolkovskiy's work enjoyed broad study in the 1950s and 1960s, as Americans sought to understand how the Soviet Union had accomplished such unexpected success in its early efforts in spaceflight. American space scientists realized upon reading Tsiolkovskiy that his theoretical efforts had provided an important basis for the development of the practical rocketry that underlay the Soviet space program.

Hermann Oberth (1894–1989) wrote the 1923 classic *Die Rakete zu den Planetenräumen* (By rocket to space). Among other important observations, Oberth suggested that space stations would be necessary for human travel to other planets. Oberth's book proved enormously important in inspiring others, including Wernher von Braun, to pursue the technological challenge of spaceflight. (NASA photo)

The German School of Space Stations

During the 1920s, Romanian-German spaceflight theorist Hermann Oberth and Austrian engineer Hermann Noordung both elaborated on the concept of the orbital space station as a base for voyages. Oberth was one of the most significant rocketry pioneers of the twentieth century. By birth a Romanian but by nationality a German—he was born on June 25, 1894, in Hermannstadt, Transylvania—as a boy of only eleven, Oberth became mesmerized by Jules Verne's novel *From the Earth to the Moon.* He recalled reading the book "five or six times and, finally, knew it by heart."[18] This book, and other spaceflight literature that he devoured in the coming years, led Oberth to intensive study of the technical aspects of interplanetary travel. Although he studied

Hermann Noordung (1892–1929) wrote *Das Problem der Befahrung des Weltraums* (1929), one of the classic early works about spaceflight. His real name was Herman Potočnik, an obscure engineer in the Austrian army who, inspired by the work of Hermann Oberth, prepared the first detailed technical designs of a space station. (NASA photo)

for a career in medicine, Oberth never could shake his obsession with spaceflight, and finally he switched his emphasis to physics. He wrote a dissertation on the problem of rocket-powered flight, but the University of Heidelberg rejected it in 1922 for being too speculative. This dissertation, however, became the basis for the classic 1923 book *Die Rakete zu den Planetenräumen* (By rocket to space). The book explicated the mathematical theory of rocketry, applied it to possible designs for practical rockets, and considered the potential of space stations and human travel to other planets.[19] Hermann Oberth, championing the cause of a space station, asserted that it would be necessary as a "springboard" for flights to the Moon.[20]

The 1923 book inspired a number of new rocket clubs to spring up all over Germany, as hardcore enthusiasts tried to translate Oberth's theories into practical space vehicles. The most important of these was the Verein fur Raumschiffarht (Rocket society), or VfR. Oberth became something of a godfather for the VfR during the 1920s, encouraging the efforts of Max Valier, Willy Ley, and the young Wernher von Braun. In 1929 Oberth published another major work, *Wege zur Raumschiffahrt* (The road to space travel), in which he envisioned the development of ion propulsion and electric rockets. He expanded on his space station concept to include not only low Earth orbit "base camp" stations but also geosynchronous space stations that could be used for astronomical observations, and polar-orbiting "strategic" stations from which reconnaissance might take place and weapons of mass destruction could be launched at any point on the globe. *Wege zur Raumschiffahrt* won an award established by the French rocket pioneer Robert Esnault-Pelterie, and Oberth used the prize money to buy rocket motors for the VfR.[21]

Oberth's idea of a space station was inspired not only by the stories of Jules Verne but also by other science fiction. As a child he had read Kurd Lasswitz's *Auf Zwei Planeten* (On two planets, 1897), which envisioned a rocket-propelled spacecraft en route to Mars. For the young Transylvanian, and for many who followed him, traveling to the red planet became sufficient cause for developing rocket technology. The first step to the stars was to enable travel to outer space, the second to build a base camp, and the third to travel from there. Oberth called it a *Weltraumbahnhof,* meaning a (small) refueling station in outer space. He liked to think of it as analogous to a coaling station, which allowed trains to travel across Europe without having to carry all of their own fuel. The space station played a key role in his plans for exploring the solar system.[22]

In later versions of the 1923 book, Oberth cloaked his space station concept in terms that made it seem a terminus justified on its own merits, regardless of how it was used. The idea of human occupancy of space he called *Aufenthalt,* an ambiguous term that could mean a temporary stop or a long-term stay. The space station was still a means to an end, but that end did not necessarily have to be the exploration of other planets. Other objectives included profit, military power, and the advancement of scientific knowledge.

From the 1920s until 1938, Oberth was involved in a series of research projects concerning rockets for the German government, some for the VfR, and later for the German Army. In 1941 he became a naturalized German citizen, and during World War II he worked for Wernher von Braun in the V-2 development program but never held an important position in the project. At the end of the war, Oberth was interrogated by American captors and released. He settled in Feucht, West Germany, near Nuremberg.[23]

Although he never gave up on the idea of a space station, Oberth had to put it on hold during the World War II era and the first years of the space age. In 1955 Wernher von Braun, by this time the head of a U.S. Army ballistic missile effort at Huntsville, Alabama, invited Oberth to work for him on the program. For a short time he did so, but in 1959 Oberth retired and returned to Feucht, where he lived the rest of his life. Appropriately, because of his significance as a godfather of spaceflight, Oberth returned to the United States in July 1969 to witness the launch of the Saturn V rocket that carried the Apollo 11 crew on the first lunar landing mission. He then returned to Germany, where he died on December 29, 1989, having helped to create and to sustain spaceflight and having witnessed many of the major events of space exploration in the latter half of the twentieth century.[24]

Although he never became the dean of spaceflight the way Oberth did, Herman Potočnik provided the most detailed technical description of a space station that appeared before the 1950s. Writing under the pen name Hermann Noordung, in 1929 Potočnik published *Das Problem der Befahrung des Weltraums* (The problem of space travel).[25] The book quickly became one of the classic writings about spaceflight. Herman Potočnik was an obscure former captain in the Austrian Navy, born on December 22, 1892, in Pola (later Pula). This was the chief Austrian-Hungarian naval station, located on the Adriatic Sea in what is today Croatia. As the location might suggest in part, his father served in the navy as a staff medical officer. The name *Potočnik* is Slovenian, also the nationality of Potočnik's mother, who had some Czechoslovakian ancestors as well. Her son was educated in various places in the Habsburg monarchy, attending an elementary school in Marburg (later Maribor) in what is today Slovenia. For his intermediate and secondary schooling, he enrolled in military schools with emphases on science and mathematics and languages, eventually earning a degree and a military commission at the technical military academy in Mödling, southwest of Vienna.

From 1918 to 1922, Potočnik studied electrical engineering at the Technical Institute in Vienna, although tuberculosis forced him to leave the army in 1919. Although he appeared to have set up a practice as an engineer, his illness evidently prevented him from working in that capacity. But he did become interested in the spaceflight movement. He contributed monetarily to the journal of the VfR, *Die Rakete* (The rocket), begun in 1927, and he corresponded with Hermann Oberth, whose 1923 book inspired him. Another Potočnik correspondent was Baron Guido von Pirquet, who wrote a series of articles on interplanetary travel routes in *Die Rakete* during 1928. He proposed space stations as depots for supplying fuel and other necessaries to interplanetary rockets. Von Pirquet's rockets, in his conception, would be launched from space stations rather than from the Earth to avoid the amounts of propellant required for escape from the home planet's gravitational field, which would be much weaker at the distance of a few hundred miles.[26]

Potočnik's book, *Befahrung des Weltraums: Der Raketenmotor* (The problem of space travel: the rocket motor), was published in 1929, the same year as his death. It deals with a broad range of topics relating to space travel, although the rocket motor mentioned in the subtitle is not especially prominent among them. What makes the book especially important in the early literature about space travel is its extensive treatment of the engineering aspects of a space station. Potočnik was hardly the first person to write about this subject, but his conceptions of a space station offered engineering details about how one might be constructed, with more than 100 engineering diagrams and even full-color illustrations.[27] Potočnik/Noordung envisioned a station located at a 23,000-mile orbit, and he speculated on the possibilities of orbits at different distances from the Earth and at other inclinations than the plane of the equator.

Noordung's vision of a space station involved three interdependent modules located in geosynchronous Earth orbit. The first was the *Wohnrad,* or habitation wheel. This part of the station followed closely the concepts of Tsiolkovskiy and Oberth for a giant rotating craft whose centrifugal force created an artificial gravity for the inhabitants. It would be constructed in small sections on Earth and then launched into orbit, where astronauts would assemble it. It would be held together by a set of spokes, looking very much like an oversize bicycle tire. Approximately 100 feet in diameter, this component of the station would be by far the largest. In the habitation wheel, Noordung wrote, "a man-made gravitational state is continually maintained

Full view of Noordung's habitat wheel for his space station as it appeared facing the sun in his 1929 book. The center concave mirror was for collecting sunlight used to generate power. The habitat was only one of three major components of Noordung's space station. The others were the observatory and the machine room, both connected to the habitat by umbilicals but not at docking ports. (NASA photo)

through rotation, thus offering the same living conditions as exist on Earth; it is used for relaxing and for the normal life function."[28]

Noordung proposed a central airlock on one side of the center of the wheel and a "solar concentrator" on the other. The latter, Noordung's name for the modern solar array, would be used to heat liquid nitrogen that circulated the entire wheel and powered turbines that generated electricity. Fearing that this would be insufficient to supply the electricity needed by the station, Noordung added a ring of solar concentrators to the wheel itself. As the station's wheel rotated to create a form of gravity, the airlock and solar concentrator at the center remained stable relative to the Earth. Noordung believed this was necessary to assist spacewalking astronauts and vehicles in using the airlock. He also linked the habitation wheel to the other two parts of the station via flexible tubes that carried the liquid nitrogen. He needed a stable place to make this connection, and the nonrotating part of the station provided this capability. Access between the microgravity environment of the nonrotating inner portion of the habitation wheel and the rotating portion would be effected by two elevators and two helical corridors.[29]

The second component of Noordung's plan was a research station, called an *Observatorium,* which was really a more generic stable, nonrotating site where microgravity experiments and scientific observations could be undertaken. Noordung believed it would include at least one telescope free from the Earth's atmosphere and light, where astronomers would be able to make the most sophisticated studies of the universe yet endeavored. It also would include advanced laboratories for materials and for biological and physical experimentation. He wrote: "The observatory contains the following equipment: primarily, remote observation equipment in accordance with the intended purpose of this unit and, furthermore, all controls necessary for remote observations, like those needed for the space mirror. . . . Finally, a laboratory for performing experiments in the weightless state is also located in the observatory."

Because this component of the station was the most important, and the raison d'être for the whole enterprise, Noordung proposed that it not be entirely dependent upon the habitation wheel's solar concentrators for its power. Instead he envisioned a third component, the machine room *(Maschinenhaus),* to power the *Observatorium.* Noordung wrote:

> The machine room is designed for housing the major mechanical and electrical systems common to the entire space station, in particular those that serve for the large-scale utilization of the sun's radiation. Primarily, it contains the main solar power plant including storage batteries. Furthermore, all of the equipment in the large transmission station is located here, and finally, there is a ventilation system, which simultaneously supplies the observatory.

Collecting solar energy takes place through a huge concave mirror firmly connected to the machine room . . . in whose focal point the evaporating and heating pipes are located, while the condenser and cooling pipes are attached to its back side. The orientation of the machine room is, therefore, determined beforehand: the concave mirror must always squarely face the sun.[30]

Astronauts during spacewalks could only accomplish movement between these three components, and neither the *Observatorium* nor the *Maschinenhaus* had facilities for food preparation or personal hygiene. They were intended for human servicing from the habitation wheel but not as a place for human occupation.

Noordung proposed using the space station for a myriad of important tasks that could only take place in Earth orbit. These fell into four major areas:

1. Scientific: ranging from what is referred to as Earth remote sensing, to astronomical observation, to microgravity materials research and processing
2. Military: strategic reconnaissance and control of weapons on the Earth's surface
3. Commercial: point-to-point communications and such unusual ideas as iceberg detection and removal via space mirrors (not so strange a concept when the Titanic was still much on the public's mind)
4. Exploratory: base camp to the stars, as the station served as an outpost on the space frontier

Noordung used these distinct elements to answer what was for him the most important question about spaceflight: "What benefits could the described space station bring mankind?"

His answers represented a good catalog of possibilities. He found that these included physical and chemical experiments conducted in the absence of gravity and heat, studies of cosmic rays, astronomical research without the interference produced by Earth's atmosphere, observations of planet Earth itself from the vantage point of space (including meteorological and military applications of the resulting information), the use of a space mirror to focus Sun rays on the Earth for a variety of purposes (including combat), and use as a base for traveling farther into space. "Furthermore," he wrote:

[I]t would be possible to generate extremely low temperatures not only in a simpler fashion than on the Earth, but absolute zero could also be approached much more closely than has been possible in our refrigeration laboratories—to date, approximately 1° absolute, that is, −272° Celsius, has been attained there—because, besides the technique of helium liq-

uefaction already in use for this purpose, the possibility of a very extensive cooling by radiating into empty space would be available on the space station.[31]

Noordung's ideas were not universally accepted, and his death from tuberculosis a few months after his book appeared meant that no meaningful dialogue over its ideas could take place. Both Willy Ley and Guido von Pirquet criticized the book roundly. Ley, for instance, recalled that Noordung had fatally erred in his insistence that the space station be placed in geosynchronous orbit. Ley wrote that

> a 24-hour orbit . . . would decrease its value by, say, 75 per cent. In such an orbit it could observe only one hemisphere of the earth and that one not very well because of the great distance, which also would make the station's construction and maintenance rather expensive in terms of extra tons of fuel consumed [on the trip to the station]. He did have a number of interesting ideas, but each one of them came out somewhat flawed.[32]

Although such criticisms served to reduce the influence of Noordung's detailed designs, the book represented original thinking on the subject and had an important place among German spaceflight enthusiasts.

It even made its way to the United States, and a very early, partial English translation by Francis M. Currier appeared serially in mid-1929 in what might seem a strange place—Hugo Gernsback's *Science Wonder Stories*.[33] There was also a translation completed for the British Interplanetary Society and kept at its library in London, and a Russian translation appeared about 1935.[34] It is possible that Noordung's work informed a short story titled "Lunetta" that Werner von Braun wrote in 1929, describing a trip to a space station. Later, of course, von Braun wrote an influential article in *Collier's* magazine in 1952 that described a space station at least superficially similar to Noordung's.[35]

Even if Noordung did not directly influence von Braun, his illustrations especially proved a powerful force on spaceflight theory before the 1960s. As just one specific example, in 1948–49

Noordung's depiction of the complete space station with its three units, seen through the door of a spaceship. In the distance is the Earth, some 26,000 miles distant. In Noordung's 1929 illustration, the station hovers in geosynchronous orbit on the meridian of Berlin, approximately above the southern tip of Cameroon. (NASA photo)

Harry E. Ross of the British Interplanetary Society, in conjunction with Ralph A. Smith, proposed a large, rotating space station based on Noordung's drawings, even though neither of them could read his German.[36]

Even more important than this direct linkage with later space station concepts, as historian J. D. Hunley concluded:

> It is also quite possible that by proposing the first actual design for a space station and by offering illustrations of that design, Potočnik helped to fixate the imagination of people interested in spaceflight upon a space station as an important goal in itself and means to the end of interplanetary flight. Since 1959 NASA has conducted at least 100 studies of space station designs, and the idea of a space station became a firm fixture in NASA's planning from the mid-1960s to the present day. Much more continuously than the United States, the former Soviet Union and Russia have had a space station program dating back to the launch of Salyut 1 on April 19, 1971. In view of the mid-1930s translation of Potočnik's book into Russian, perhaps that program, too, owes something to the little-known Austro-Hungarian engineer's study.[37]

This conclusion has been documented in numerous historical studies, and in this context Herman Potočnik, a.k.a. Hermann Noordung, has cast a long shadow over space station development concepts since publication of his book in 1929.[38]

One would be hard pressed to find serious spaceflight theorists of the interwar period, or even writers of sophisticated science fiction, who did not envision a space station as a necessary step in the process of humanity leaving Earth. Space stations took on the persona of base camp to the stars in an overt manner during this time, aided in a most essential way by the work of such writers as Hermann Noordung. In 1944 Willy Ley expressed well the conventional wisdom then in existence about the significance of a space station. "The realization of the station in space is the realization of space travel in general," Ley announced. "Trips to the moon, around the moon, and even to the other planets are no longer difficult if they are made from that station."[39]

But it was after World War II that the concept of a giant, rotating station reached its zenith. Conceived by rocket engineer Wernher von Braun after his immigration to the United States in 1945 and visualized by space artist Chesley Bonestell, that station solidified in many Americans' minds a great wheeled permanent presence in space where many lived and worked, and from which humanity ventured on to other celestial bodies.

Many of the designs for space stations during the interwar period were quite exotic and highly impractical. Here, in violation of the laws of physics, space planes travel to and from a docking station in Earth orbit while the planet passes below. (NASA photo)

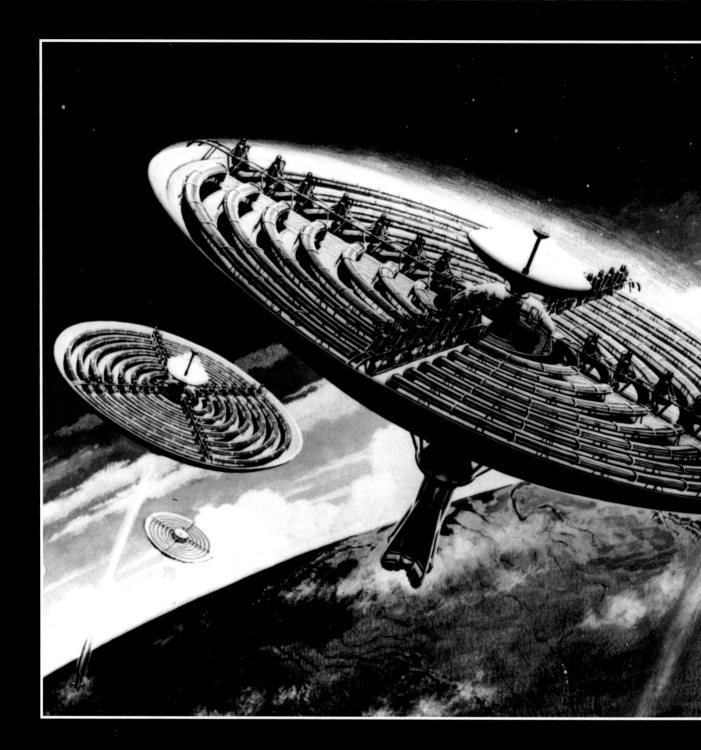

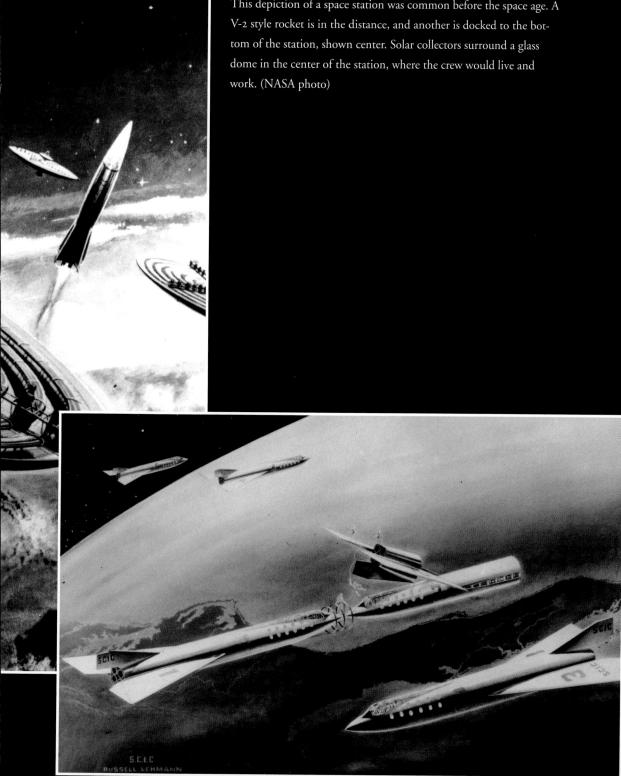

This depiction of a space station was common before the space age. A V-2 style rocket is in the distance, and another is docked to the bottom of the station, shown center. Solar collectors surround a glass dome in the center of the station, where the crew would live and work. (NASA photo)

S.C.1.C
RUSSELL LEHMANN

CHAPTER 2

The Space Station and the von Braun Paradigm

During the period between the early 1950s and the early 1970s, no single individual dominated the space policy debate more thoroughly than Wernher von Braun, the mastermind of the German V-2 during World War II. After his immigration to the United States in 1945, he soon became the most eloquent spokesperson for space exploration in the United States. A key ingredient of his popularity rested with his promotion of an integrated space exploration agenda that was at once ambitious and logical, visionary and compelling. Vital to his strategy was a space station, modeled after the earlier efforts of Konstantin Tsiolkovskiy, Hermann Oberth, and Hermann Noordung. Von Braun effectively espoused the vision of a mighty wheel-like space station that housed at least fifty inhabitants, serving as a scientific and military platform for Earth-oriented activities and as a base camp to the stars beyond. It floated in orbit Janus-like, simultaneously looking back at Earth—where it might provide a myriad of services—and outward to the Moon and planets, where it served as the departure point for human exploration and colonization.

Through a strenuous schedule of speaking and writing, von Braun urged his vision on the rest of the United States, inspiring many to adopt his ideas. Teenager Homer Hickam, for one, idolized von Braun in the aftermath of the launch of Sputnik 1 in October 1957. The boy saw him as the one person in the United States who had both an imagination and a plan sufficient to create a constructive future in space. Von Braun's positive vision for an expansion of humanity beyond Earth stirred young Hickam, growing up in a coal-mining town in West Virginia far removed from the mainstream of spaceflight doctrine, to master mathematics and science and, later, to go on to a career as a NASA engineer.[1] Over time, von Braun's integrated space exploration strategy has come to be called the "von Braun paradigm."[2]

The von Braun Paradigm

In the 1950s, all of the previous rationales for a space station and the destiny of human travel to other planets found expression in a public-relations campaign mounted in the United States by Wernher von Braun, a handsome German émigré and charismatic leader. He assaulted the print and electronic media with a compelling concept that has become *the* grand plan for human spaceflight ever since. Von Braun's reputation preceded him; he had masterminded the V-2 ballistic missile of Adolph Hitler's Germany during World War II. Developed in a secret laboratory at Peenemünde on the Baltic coast, this rocket may be viewed as the immediate antecedent of some of those used in the U.S. space program. At war's end, von Braun deliberately surrendered to the Americans in the hope that he and his key technical people could immigrate to the United States and work on rockets there.[3]

His mission accomplished, von Braun called for a scenario centered on human movement beyond this planet and involving these basic ingredients, accomplished in essentially this order:

1. Launch Earth-orbital satellites to learn about the requirements for space technology that must operate in a hostile environment
2. Begin Earth-orbital flights by humans to determine whether it is really possible for people to explore and settle other places
3. Develop a reusable spacecraft for travel to and from Earth orbit, thereby extending the principles of atmospheric flight into space and enabling routine space operations

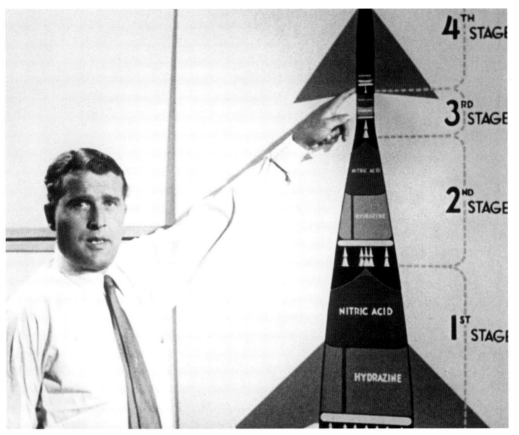

Wernher von Braun (1912–1977) was perhaps the most effective proponent of a space station as a base camp to the stars. He appeared in three Walt Disney television specials to advocate his integrated space exploration agenda, often called the von Braun paradigm. Here he is seen explaining the space plane that will travel from Earth to a space station in Disney's production "The Challenge of Space," in 1957. (NASA photo)

4. Build a permanently inhabited space station as a place both to observe the Earth and from which to launch future expeditions (this serves as the base camp at the bottom of the mountain, or the fort in the wilderness, from which exploring parties depart)
5. Undertake human exploration of the Moon, with the intention of creating Moon bases and eventually permanent colonies
6. Begin human expeditions to Mars and eventually colonize the planet

This has become known over time as the von Braun paradigm for human colonization of the solar system. Von Braun espoused these ideas in a series of important *Collier's* articles over a three-year period in the early 1950s, each with striking images by some of the best artists of the era, and later in a classic set of three Disney television programs.[4]

The focus of von Braun's provocative piece in the first *Collier's* issue devoted to spaceflight was an imaginative design for a human-operated space station placed permanently in Earth orbit. In the article, "Crossing the Last Frontier," von Braun wrote, "Development of the space station is as inevitable as the rising sun." He added: "Man has already poked his nose into space" with sounding rockets, and "he is not likely to pull it back. Within the next 10 to 15 years, the earth will have a new companion in the skies." An "artificial moon," an Earth-orbiting base "from which a trip to the moon itself will be just a step," will be "carried into space, piece by piece, by rocket ships." From there, human civilization headed for deep space will embark.[5]

The *Collier's* series catapulted von Braun into the public spotlight like none of his previous scholarly activities had been able to do. The magazine was one of the four highest-circulation periodicals in the United States during the early 1950s, with more than 3 million copies sold each week. If readership was indeed four or five people per copy, as the magazine estimated, something on the order of 15 million people were exposed to these spaceflight ideas. *Collier's*, seeing that it had a potential blockbuster, did its part by hyping the series, with advertisements of the space artwork that appeared in the magazine, more than 12,000 press releases, and media kits. It set up interviews on radio and television for von Braun and the other space writers—but especially von Braun, whose natural charisma and enthusiasm for spaceflight translated well through that medium. He appeared on NBC's "Camel News Caravan" with John Cameron Swayze; NBC's "Today" show, then hosted by Dave Garroway; and CBS's "Gary Moore" program. Whereas *Collier's* was interested in selling magazines with these public appearances, von Braun was interested in selling to the public his concept of sustained space exploration.[6]

Following close on the heels of the *Collier's* stories, Walt Disney Productions contacted von Braun—through science writer Willy Ley—and asked for his assistance with three shows for Disney's weekly television series. The first of these, "Man in Space," premiered on March 9, 1955, with an estimated audience of 42 million. The second show, "Man and the Moon," also aired in 1955 and sported the powerful image of von Braun's wheel-like space station as a launching point for a mission to the Moon. The final production, "Mars and Beyond," premiered on December 4, 1957, after the launching of Sputnik 1. Von Braun appeared in all three shows to explain his concepts for human spaceflight, while Disney's characteristic animation illustrated the basic principles and ideas with wit and humor. Of course, each of the shows depicted the concept for a space station evolving from earlier ideas into von Braun's unique conception.[7]

Although some scientists and engineers criticized von Braun for his blatant promotion of both spaceflight and himself, the *Collier's* series and especially the Disney programs were exceptionally important in changing public attitudes toward spaceflight. Media observers noted

Hermann Oberth *(left)* and Wernher von Braun *(center)* listen to a presentation on orbital trajectories by Charles Lundquist, at the Army Ballistic Missile Agency in Huntsville, Alabama, June 28, 1958. (NASA photo)

the favorable response to the three Disney shows and recognized that "the thinking of the best scientific minds working on space projects today" had gone into them, "making the picture[s] more fact than fantasy."[8]

At a fundamental level, Wernher von Braun played upon the centuries of American experience in exploration and settlement and the importance of forts on the frontiers of the American West. Terrestrial forts and bases, he commented, became a conventional method for advancing human presence into unconventional territory. Some became legendary—Fort Pitt, Fort Laramie, Fort Leavenworth, and Bent's Fort—others gave their names to the cities that grew up around them, such as Fort Still in Oklahoma and Fort Walton Beach in Florida. All served an important purpose and have enjoyed a positive image in mainstream American culture.[9] As historian Howard E. McCurdy concluded: "Colonists and pioneers, mountain climbers, and polar explorers commonly employed forts and base camps as a means of marshalling material and fashioning sanctuaries in hostile realms. By the time that the space age began, the image of the frontier fort or

Chesley Bonestell's 1952 panorama of a giant wheeled space station presented a powerful icon for American society of what a space station should look like. As it floats in low Earth orbit above Central America, a crew of spacewalking astronauts unload cargo from a winged shuttle. A free-flying observatory is also being tended by astronauts in the foreground. (Courtesy of Bonestell Space Art)

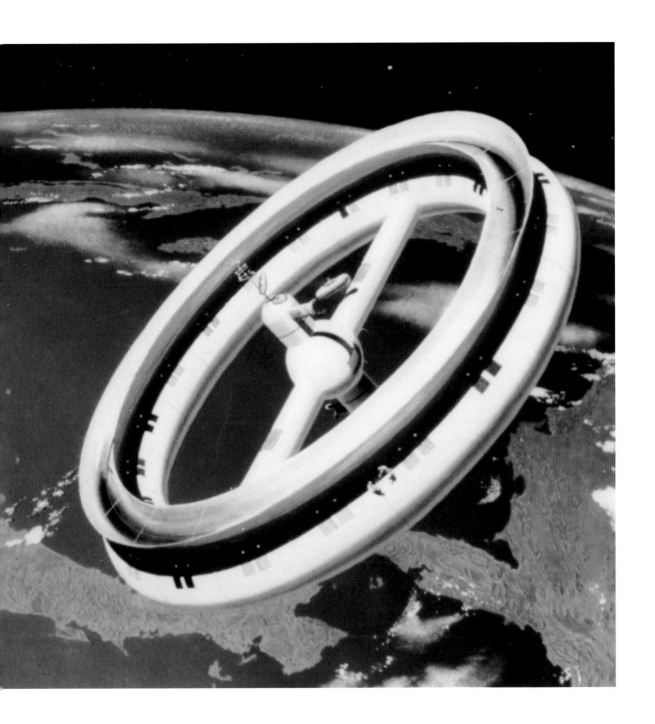

base camp as a jumping off point for more daunting adventures was well established in the public mind."[10] It was certainly appropriate for Americans to view space stations in the same manner.

Von Braun played to a very specific American self-image, one in which frontier forts and way stations dotted the landscape. McCurdy observed:

> From the garrison houses of New England to the station settlements of Kentucky and Tennessee, from the brick and earthen fortresses of the Atlantic coast to the simple stockaded trading posts of the West, Americans built forts and stations as an initial step in settling unfamiliar territory. . . . Frontier forts offered safe havens from the dangers of the wilderness, attracted commerce and trade, and provided jumping off points for further ventures into the wilderness.[11]

The importance of the westering experience in American history provided a fruitful ground for accepting von Braun's concepts. In the early 1900s, the fort-in-the-wilderness concept found another expression in the base camps of polar explorers and mountain climbers, and von Braun expanded further on the analogy.[12]

He explained his full vision of a space station in the March 1952 issue of *Collier's,* suggesting that giant rockets weighing 7,000 tons each would place 36 tons of material in space with each flight, which would allow the construction of a large station that would house a crew of up to eighty astronauts. They would assemble three large spacecraft that would be used on the first voyage to the Moon. Later flights would place an increasing amount of equipment and supplies in Earth orbit, facilitating the eventual establishment of a lunar base.[13]

Von Braun proposed that the station measure 250 feet wide and orbit the poles at 1,075 miles above the Earth's surface. Shaped like a wheel, the station would make a complete turn on its axis every twenty-two seconds, centrifugal force simulating one-third gravity at the outer rim, where most of the inhabitants would reside. Artist Chesley Bonestell painted a splendid picture to accompany the article: a wheel-shaped station with graceful lines passes over Central America, with the Gulf of Mexico, Mexico, Cuba, and the Florida peninsula in the distance. Although sparse cloud formations dot the landscape in the distance (far fewer than may actually be found on any given day), from orbit the Earth appears lush and green, almost serene. Astronauts in space suits work in a nearby observatory, while others unload supplies from a winged shuttle recently arrived from Earth. Space taxis shaped like overgrown oval aspirin pills carry astronauts among the various entities in orbit. With this painting, Bonestell produced one of the most enduring images of a space station yet created.[14] It would inform nearly every conception of what a space station should look like for the next quarter-century.

To illustrate a separate article by Willy Ley in the same issue of *Collier's*, artist Fred Freeman added to the Bonestell vision by painting a cutaway of one section of the space station wheel. The space station crew works on three decks inside the outer rim, much like sailors on a nuclear submarine. Elevators carry workers to the station hub; regulators and pipes keep it cool. Von Braun's method of power generation found its way into both Freeman's and Bonestell's illustrations. The former V-2 mastermind estimated that the space station would require 500 kilowatts of electricity, which he proposed to achieve through a condensing mirror and generator. A highly polished metal trough would run the circumference of the wheel, concentrating the rays of the Sun onto a steel pipe containing liquid mercury. This would transform the mercury into hot vapor, which in turn would drive a turbogenerator at the axis of the wheel.

To simplify construction of this space station, von Braun suggested that it be fabricated out of inflatable nylon and plastic. The material could be collapsed for its trip from the Earth, then pumped up once in orbit. Twenty sections, each an independent unit, would be assembled to complete the wheel. This set the stage for an entire genre of concepts in the late 1950s and 1960s, with space stations looking more like automobile tires than even the designers wished. Constructing and supplying the orbital facility would require frequent resupply missions. Von Braun wrote, "There will nearly always be one or two rocket ships unloading supplies."[15]

Von Braun also envisioned a space station as a representation of national sovereignty in Earth orbit. Accordingly, the space station held the same promise as overseas bases for the U.S. Navy. Territorial claims to celestial bodies would need to be extended, von Braun realized, and a station would help legitimate these. He thought of the Moon and the planets in the same way that the Europeans thought of the Western Hemisphere, a place to be colonized and subjugated. The space station would serve a valuable purpose in this effort. In addition, the editors of *Collier's* declared, "Whoever is first to build a station in space can prevent any other nation from doing likewise."[16]

The von Braun space station immediately captured the enthusiasm of advocates. The powerful image of the rotating station became the one that entered the popular imagination. From that point on, it has found itself repeated in art, literature, and film.

The importance that people attached to his vision of a space station embarrassed von Braun at least a little. Privately, he dismissed the *Collier's* station as a "monster" that had assumed importance beyond what he had originally intended. Von Braun believed the station was but one small part of a much larger scheme, part of a host of related technologies that included rockets, guidance and navigation systems, life support systems, space suits, in-space maneuvering rockets (or "reaction control systems"), refueling tanks, and so on. He remarked that all of this needed to be developed first in small spacecraft before grander schemes, such as a space station, could

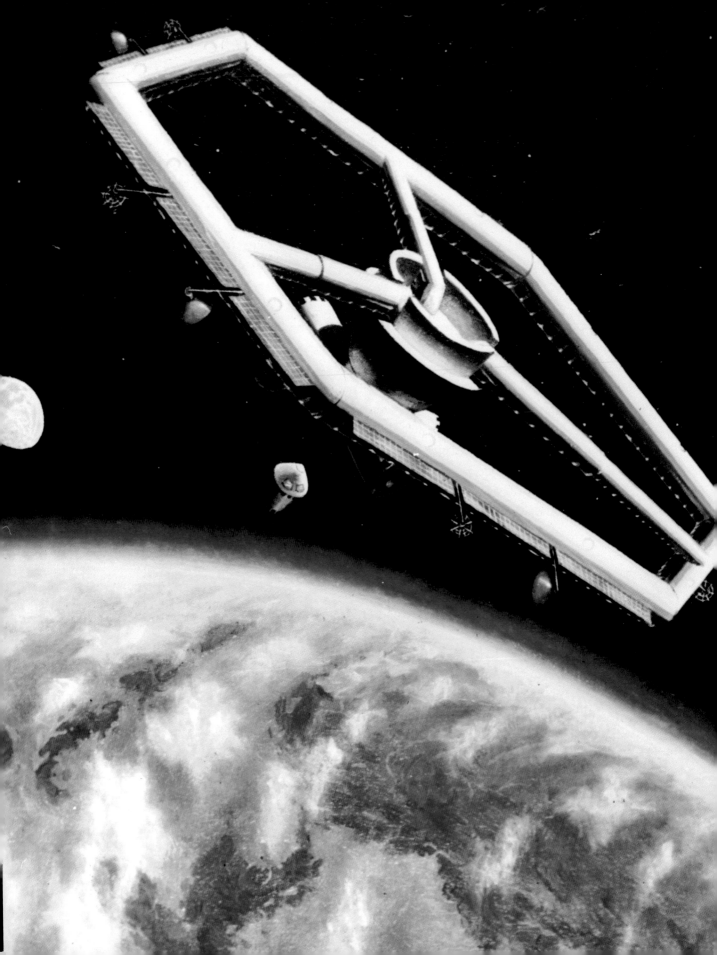

even be considered.[17] But publicly, von Braun remained firmly committed to the big station as a critical step in his integrated space exploration plan.

SUCCESSORS OF THE VON BRAUN STATION

Wernher von Braun's sober analysis of the need for a space station touched off a craze of design efforts in the 1950s from several quarters. There were seemingly as many blueprints for space stations as there were engineers eager to draw them. Most never were seriously considered and remained in the files of the individual designers. Some were exotic, others impractical, and a few represented works of genius. Three engineers were in this latter category: Heinz H. Koelle, Darrell C. Romick, and Krafft A. Ehricke. All created elaborate space station designs, in the process clearly showing the relationship between technical design and construction costs.

Koelle, who came to work for Wernher von Braun after 1955 as head of his Future Projects Office, was a young German who first learned of von Braun's ideas from the book *Das Mars Projekt*, published in Germany in 1947. They entered into a lively correspondence while Koelle was a student at the Technical University at Stuttgart in 1950. Koelle recognized, perhaps more than even von Braun, that the cost of a space station would be enormous, not only for construction but for all operations. He argued that it had to be a civilian project, rather than a military one, the purposes of which must be oriented toward a laboratory to prepare for "extensive space travel projects," a base to realize those projects, a scientific research laboratory, and an instrument of international prestige. Space enthusiasts could rely upon the deep coffers of the military to fund the effort, he noted, but to obtain civilian sector support, always more sparse than funding from the military, Koelle demanded both that space station designs be as economical as possible and that they be as nonthreatening as an overhead platform could be.[18]

Koelle believed that the cost of assembling a station in orbit would be the single highest expenditure for any space station program, and he made a major contribution by designing prefabricated modules that could be mated—rather than assembled—on orbit. Launched as an upper stage of rockets then in development, Koelle envisioned a space station connected in microgravity by astronauts, with a minimum of work. He anticipated a strategy that would be used by both the United States for the Skylab program and the Soviet Union for a series of Salyut space stations. Indeed, in the 1990s the International Space Station employed this basic strategy of building on the ground and then launching and plugging the modules together on orbit.[19]

Darrell C. Romick, a young aeronautical engineer at the Goodyear Aircraft Corporation, had

The wheeled space station of von Braun inspired many imitators. This concept depicts a hexagonal station that could provide artificial gravity through centrifugal force. (NASA photo, no. Space-Ship-13)

similar but independent ideas. In 1954 he began work on a three-stage "ferry rocket" to serve as the basic building block for a space station, with the upper stage containing a station module. Emphasizing six design parameters—immediate operation, continuous operation, growth and expansion in a process of evolution, maximum payload for minimum launches, fast and simple construction, and safety achieved through compartmentalization—Romick sought to define the most efficient, least costly space station design.[20] He urged an incremental approach, essentially buying a station "by the yard," by building self-contained modules launched and placed into operation even as the wheel-like structure took shape. In Romick's vision, a module already on orbit would house a crew that could begin immediately to conduct research. As additional modules were launched from Earth, the astronauts aboard the already inhabited part of the station could plug them into the operational modules. Indeed, if each building block was fully outfitted and had separate living quarters, space station crews could begin work immediately instead of waiting for additional components. This strategy provided an added advantage. In the event of an accident in one of the compartments, the others would provide a safe haven. Romick's ideas did not receive much consideration at the time—many did not even understand the implications—but his approach for turning a small space station into a large one found expression in the building of the ISS.[21]

Krafft A. Ehricke was the third aerospace engineer to contribute significantly to the quest for a space station in the 1950s. One of von Braun's team from Peenemünde, Ehricke had joined von Braun at Fort Bliss, Texas, in 1947, but had moved on to work for the Convair division of General Dynamics in 1954. There he helped with design work for the U.S. Air Force's Atlas program. Similar to von Braun, Ehricke was a public intellectual, once called "one of the world's foremost apostles of space travel."[22] Ehricke advocated a small, single-purpose station, in contrast to the big-wheeled variety, because he realized that the larger version's price would be too high to get any nation to fund it. He doubted that the military services of the United States would fund anything beyond their own immediate interests. He recognized, in a fit of practicality not often seen among space enthusiasts, that a station, at least in the 1950s, was not necessary to national defense.

Ehricke noted that the military was interested in space planes, satellite communications, reconnaissance satellites, and ballistic missiles, something he fully appreciated, and that they had invested heavily in those technologies. As one example of governmental investment in space technology, the Dwight Eisenhower presidential administration undertook a study in the summer of 1957 and learned that through fiscal year 1957 the nation had spent $11.8 billion on military space activities. "The cost of continuing these programs from FY 1957 through FY 1963," Eisenhower was told, "would amount to approximately $36.1 billion, for a grand total of $47 billion."[23]

In 2001 dollars, for comparison, this would have represented an investment of more than $230 billion. There was no way that the U.S. government would essentially double that investment to add a space station to the list of projects. "Beyond the reconnaissance station," Ehricke wrote, "the conditions for astronautical progress will change radically, because the main customer for the products of this development will drop out."[24] Ehricke believed that further intensive investment after reconnaissance satellites would depend on what new opportunities space systems might open up.

Because of this, Ehricke advocated a flexible space station effort in which engineers tailored specific pieces of equipment to the needs of the mission and then placed the station into the most useful orbit for its purpose. For example, if the purpose of a space station was worldwide weather observation, then a "high" inclination orbit (circling the Earth around the poles) would serve better than a "low" one (circling the Earth around the equator). A small, dedicated-purpose space station would be cheaper, Ehricke rationalized, because it would not need a large crew. He wrote that a sufficient personnel number might be four, or perhaps even less.[25] His ideas had validity, but Ehricke failed to anticipate the electronics revolution that led to all kinds of robotic spacecraft that could perform the same missions he described. Such satellites essentially defeated any reason to build single-mission space stations with crews aboard.

He came closest to finding a market for his station concept in early 1958, in the aftermath of the Sputnik crisis, by proposing to the U.S. Air Force a Minimum Manned Satellite. The MMS project, sponsored by General Dynamics, called for building a station with three Atlas rockets. The empty fuel tank of the first Atlas would become the base, a second would bring supplies and equipment, and the last would contain two capsules with four astronauts aboard. They would assemble everything on orbit to create a small station.[26] The plan never had a chance. The first and insurmountable problem was how to make a spent upper stage habitable by humans. The workload required on orbit was enormous. Later, in the Skylab program, NASA would seriously consider and ultimately reject the idea of using an expended upper stage of a Saturn V as the habitat, because of the challenges of reconfiguring it for use as a space station.[27]

A closely related proposal emerged in 1959 from the Army Ballistic Missile Agency in Huntsville, Alabama, where von Braun's rocket team reigned. Project Horizon was the army's last bid for a role in human spaceflight. Ostensibly a proposal to establish and maintain a military base, the Horizon study reflected the longstanding belief that movement beyond Earth orbit required a space station as a staging location. At a minimum, although the study's authors recommended a large wheel-like station such as von Braun's as the ideal transshipment point, any vehicle destined for the Moon would need to be refueled in Earth orbit. To accomplish that task, an orbital shelter housing fuel and a crew to support refueling would be needed. The facility could be an expended third stage refitted for crew quarters, control systems, and fuel. Over several years, a total of twenty-two of these spent stages could be joined together to make a fullfledged space station of the desired wheeled shape. It is unlikely that this study had any support elsewhere in the U.S. Army, but even if it had, within a year of its preparation von Braun's team transferred to NASA, and the report went into the dusty files of the National Archives without further consideration.[28]

So the idea of the large, multipurpose space station remained, albeit one that could be constructed using a building block approach and thereby at a somewhat reduced price. But the one aspect that neither von Braun nor his successors ever mastered—and this was especially true of Project Horizon—was a compelling rationale for building a huge and costly station. Of course it would be a laboratory in space, a scientific research facility, an observatory, a drydock for space repair, and a base camp to the stars, but none of these missions had the capacity to justify the high price tag. Only two possibilities could even begin to do so.

The first was a military mission. The people of the United States would willingly spend for national defense, and at the height of the Cold War the need to develop new technologies was readily apparent. But Ehricke was right: at no time in the 1950s did the military foresee a necessity for any type of space station. The other possibility was based on national prestige, essentially telling the world through the building of a space station that the United States was the leader in space technology. Wernher von Braun explicitly made this case throughout the decade, seeking to sell a station to the public on the basis of status rather than function. "Utility justifications of space flight have always sounded very artificial to me," he complained to Ehricke just a month before the launch of Sputnik, "and actually, are not even in vogue in modern sales techniques any more. When all automobiles are almost equally good, you do not sell a Cadillac by pointing out its technical advantages; you sell it under the slogan that you are buying more prestige by driving a Cadillac."[29] In the end, it was the desire for international prestige that took the United States to the Moon in the 1960s. Perhaps it could be drawn upon again to yield a space station as a permanent base.[30]

PERMUTATIONS OF VON BRAUN'S SPACE STATION AT LANGLEY RESEARCH CENTER

Even as Wernher von Braun and others were exploring myriad concepts throughout the 1950s, a small team of NASA engineers and contractors working at the Langley Research Center in Hampton, Virginia, was developing very different ideas on how to build an orbital research facility. Although the Langley researchers have not received credit for their pioneering efforts, not even von Braun's research team in Huntsville was more significant than the sleepy government laboratory located among tobacco fields in the tidewater of Virginia. There, beginning in the 1950s, teams of researchers tackled the diverse problems of designing, building, and operating an effective space station.

Only a year after von Braun published the first *Collier's* article, a Langley study defined a concept for a piloted research airplane that could fly to the edge of the atmosphere and then be boosted by rockets into space. Many at Langley thought that this sort of vehicle, which would return to Earth

Not all early space station concepts were for wheels, but they still sought to create artificial gravity. Here is an early design resurrected in 1969, in the space station planning effort of the post-Apollo era. It depicts a station assembled from spent stages from the Apollo program, assembled on-orbit into an oddly configured and quite impractical station that rotated on its central axis to provide artificial gravity. (NASA photo, no. S-69-1635)

by gliding under aerodynamic control, could serve as a shuttle that might eventually service an orbiting laboratory by moving personnel and supplies back and forth between Earth and space. Beyond this sort of speculation, however, no specific or formal thinking for the design of a space station took place at Langley until after Sputnik and the birth of NASA.[31]

At that point, however, serious thinking about a space station took off. As one example, by the end of 1959—little more than a year after NASA had begun operations—the agency had prepared a formal long-range plan that recognized a space station as a base camp. In addition to other objectives for the 1960s, the plan called for the "first launching in a program leading to manned circumlunar flight and to a permanent near-earth space station" in the 1965–67 period, and then using that station as a base camp for reaching the Moon sometime "beyond 1970."[32] Although the first NASA long-range plan featured a balanced program of science, applications, and human spaceflight, from the start it was the promise of flying in space that excited the public and that NASA, therefore, made the focal point of its thinking.

While this took place at headquarters, NASA's Langley Research Center led a charge to move forward with a space station. It set up preliminary working groups concerned with space station technology and encouraged the same at various firms in the aerospace industry. In early April 1959, John W. Crowley, NASA's director of Aeronautical and Space Research, set up a ten-member Research Steering Committee on Manned Space Flight. This inter-center group, chaired by Harry J. Goett of Ames Research Center, coordinated studies within NASA to ensure there was "an adequate foundation of basic research applicable to the problems of manned space flight." Langley's representatives were Laurence K. Loftin and Maxime A. Faget from NASA's Space Task Group, charged with carrying out Project Mercury. At the committee's first meeting, Loftin articulated a vision of NASA's objectives, including "manned exploration of the moon and planets and the provision of manned earth satellites for purposes of terrestrial and astronomical observation." He explicitly said that "the next major step in the direction of accomplishing these long-range objectives of manned space exploration and use would appear to involve the establishment of a manned orbiting space laboratory capable of supporting two or more men in space for a period of several weeks."[33]

These discussions led directly to an April 1960 space station symposium cosponsored by NASA, the Rand Corporation, and the Institute of Aeronautical Sciences. Various designers presented detailed concepts, but it was Lockheed Corporation's elongated, nuclear-powered version that most impressed the audience. Designed to be carried aloft by a Saturn-class booster, it consisted of modules to be assembled on orbit. Docking and scientific experiments were to be carried out in a zero-gravity module, while a habitation module featured artificial gravity (generated via rotation). Power was to come from a SNAP II nuclear-power system mounted at one end. The other end, more than 200 feet away, would have multiple docking facilities. Yet the innovative design had a key drawback: it lacked the torsional stiffness that engineers believed would be critical to stability.[34]

The results of this symposium energized several at Langley to undertake detailed feasibility studies of half a dozen potential configurations. The studies aimed to integrate four elements: operational demands for a space laboratory, technical capabilities, life support requirements, and cost constraints. Langley engineers also developed a strong conception of what they were attempting. They envisioned an orbital base for these purposes:

- Learning how to survive in space
- Undertaking scientific experiments in the absence of gravity
- Carrying out space applications research and operations, such as telecommunications and Earth observation
- Evaluating technologies for extended space exploration, including launch capabilities from the station

From these clear objectives, Langley engineers examined six different designs for a space station, all equipped with motors to rotate them for the purpose of providing artificial gravity. These well-grounded but wide-ranging studies served NASA well as reference points for any future work it might undertake.[35]

The most important concept considered in the early 1960s was some sort of self-deploying inflatable habitat. Langley engineers had embraced inflatables for communications satellites (such as the passive reflector Echo 1 balloon), and they saw no reason why such technology would not work for a space station.[36] Beyond that, the balloon concept made good engineering sense. It took hundreds of pounds of propellant to put any one pound of payload into orbit, so whatever could be done to make a payload lighter was to the good. Indeed, one of the major reasons why NASA Langley was so interested in applying the concept of inflatables to a space station was because it had by definition to be quite large in volume. Engineers wanted to "squeeze it down into a small package on top of the boost vehicle, get it into orbit, and then in some fashion allow it to blossom into its desired shape."[37]

The first serious idea for an inflatable space station to surface at Langley carried the generally unfathomable name "Erectable Torus Manned Space Laboratory." This design, developed in tandem by Langley and Goodyear Aircraft Corporation engineers, called for a flat inflatable ring or "torus" some 24 feet in diameter, or about one-quarter the size of the Echo 1 sphere. One of the major advantages of the blow-up torus was that it was a single structure that could be carried to orbit by one booster. NASA could simply fold it into a compact payload, as it was doing with the Echo balloons, for an automatic deployment once the payload reached orbit. The inner volume of the torus could generate up to 1 g through rotation; it could be designed for both natural and artificial stability, for a rendezvous-dock-abort capability and for variable-demand power supply; and it could include regenerative life support systems for a six-person crew. For electric power, the station could use a solar turboelectric system, involving an innovative umbrella-like solar collector then under development by TRW (Thompson, Ramo, and Wooldridge) as part of the NASA-supported Sunflower Auxiliary Power Plant Project.[38]

The inflatable concept was nothing short of brilliant. However, it never went very far for several distinct reasons. First, critics raised the problem of micrometeoroids slamming into the structure and causing catastrophic failure. Next, many expressed misgivings about its solidity—would it be strong enough to stand up to human occupancy for long? But mostly, skeptics thought the inflatable structure was just too inelegant. How could anything that looked like an oversize automobile tire become NASA's first space station?[39] Although technical considerations were very real in the 1960s, through years of research, by the mid-1990s most of them had been overcome. Balloon space structures reemerged as an important discussion item for future operations, including for some modules of the ISS. For instance, in 1997 NASA began work on a unique, inflatable module to provide living quarters for astronauts aboard the ISS. Called Transhab, it may, in fact, prove to be a precursor for space-based private hotels of the future.

What held more immediate promise in the early 1960s was a modular space station, as it was called by Langley engineers. The idea emerged from the mind of one of them—Rene Berglund, in the summer of 1961. It soon led to a six-month study contract with North America Aviation, Inc. Berglund's concept called for a large modular space station that, although essentially rigid in structure, could still be automatically erected in space. In essence, the idea involved putting together a series of six rigid modules that were coupled or connected by inflatable connectors coming off a central nonrotating hub. The structure, 75 feet in diameter, could be assembled entirely on the ground, packaged into a small launch configuration, and boosted into space on top of a Saturn rocket. One of the major incentives behind Berglund's design was to provide protection against micrometeoroids. To accomplish this, he gave the rigid sections of the rotating hexagon airlock-type doors that could be closed tight whenever there was any threat to the integrity of the interconnecting inflatable sections. This would enable astronauts to close off any part of the hexagon that had been struck by a micrometeoroid.[40]

This sophisticated modular assembly would rotate slowly on orbit, making it essentially a scaled-down version of von Braun's space station. In fact, a diameter of 75 feet was selected because it provided the minimal radius needed to generate a low rotational velocity that could still produce a semblance of gravity. The only part of the structure that would not rotate was the central hub; suspended by bearings, it would turn mechanically in the opposite direction of the hexagon at just the right rate to cancel out all the effects of the rotation. Located in this nonrotating center of the space station would be a laboratory, where microgravity experiments could be carried out. The nonrotating hub would also be the place for the docking of a shuttle from the Earth's surface. Preliminary experience with Langley's earliest rendezvous and docking simulators had shown that a trained pilot could land with surprising ease as long as the station-docking hub was fixed.

Convinced that they were pursuing a beneficial course, North American and Langley engineers asked for funds to begin serious design work. At that point they reached a roadblock. With the majority of NASA energy expended toward Project Apollo, no one would willingly support significant funds for a space station. By mid-1962 this proposal had ceased to have any chance of funding, especially after the decision to conduct Apollo using a lunar-orbit rendezvous approach. As long as the possibility had existed that NASA might select Earth-orbit rendezvous as the mode for reaching the Moon, a space station used as a refueling location might have gained support as part of Project Apollo. But without taking that route, one could not be supported.

Langley continued to explore the concept in studies throughout the 1960s, but without much possibility of gaining approval to build one.[41]

Von Braun and Earth-Orbit Rendezvous

Like the engineers at the Langley Research Center, Wernher von Braun believed that the Apollo program might necessitate a space station in Earth orbit. As we have seen, he felt that an Earth-orbiting space station represented a logical step in the process of exploration, and everything possible had to be done to attain one. This took the form, as NASA expanded for Apollo, of urging Earth-orbit rendezvous for the lunar missions. If selected, this option would require some type of facility in orbit that would serve as a base camp for trips to the Moon. It was a genuine possibility, in the 1961–62 period, that Apollo could be accomplished using this method. No controversy in Project Apollo more significantly caught up the tenor of competing constituencies in NASA than this one. NASA seriously considered three possible scenarios for the Moon trip:

1. Direct ascent called for the construction of a huge booster that launched a spacecraft, sent it on a course directly to the Moon, landed a large vehicle, and sent some part of it back to Earth. The Nova booster project, which was to have been capable of generating up to 40 million pounds of thrust, would have been able to accomplish this feat. Even if other factors had not impaired the possibility of direct ascent, the huge cost and technological sophistication of the Nova rocket quickly ruled out the option and resulted in its cancellation early in the 1960s, despite the conceptual simplicity of the direct ascent method. It had few advocates when serious planning for Apollo began.
2. Earth-orbit rendezvous was the logical first alternative to the direct ascent approach. This called for the launching into orbit of various modules required for the Moon trip. Above the Earth they would rendezvous, be assembled into a single system, be refueled and sent to the Moon. This could be accomplished using the Saturn launch vehicle already under development by NASA and capable of generating 7.5 million pounds of thrust. This method of reaching the Moon, however, was also fraught with challenges, notably finding methods of maneuvering and rendezvousing in space, assembling components in a weightless environment, and safely refueling spacecraft. It also added to the technological complexity because it necessitated expending precious resources to develop a station.
3. Lunar-orbit rendezvous proposed sending the entire lunar spacecraft up in one launch. It would head to the Moon, enter into orbit, and dispatch a small lander to the lunar surface.

The most unusual, intriguing, and perhaps uniquely viable concept for a space station was one that inflated after reaching Earth orbit. The design for this 24-foot, two-person space station came from the Goodyear Aircraft Corporation in 1961, and ground tests proceeded for several years thereafter without anything flightworthy being built. In the 1990s inflatables returned to the space engineering vocabulary and several concepts are currently on the drawing board. (NASA photos, nos. Misc #-12A, 62-Space Station-1)

This was the simplest of the three methods, in terms of both development and operational costs, but it was risky. Since rendezvous was taking place in lunar instead of Earth orbit, there was no room for error, or the crew could not get home. Moreover, the spacecraft would have to perform some of the trickiest course corrections and maneuvers after the crew had started a circumlunar flight. In the end, the Earth-orbit rendezvous approach kept all the options for the mission open longer than the lunar-orbit rendezvous mode.[42]

Advocates of the various approaches contended over the method, while the all-important clock of landing on the Moon by the end of the decade continued to tick. It was critical that a decision not be delayed, because the mode of flight in part dictated the spacecraft developed. Although NASA engineers could proceed with building a launch vehicle, the Saturn, and define the basic components of the spacecraft—a habitable crew compartment, a baggage car of some type, and a jettisonable service module containing propulsion and other expendable systems— they could not proceed much beyond rudimentary conceptions without a mode decision. Some inside NASA pressed hard for the lunar-orbit rendezvous as the most expeditious means of accomplishing the mission. But von Braun and his colleagues at NASA's Marshall Space Flight Center in Huntsville, Alabama, held out for Earth-orbit rendezvous.[43]

At the same time, von Braun's team was conducting studies of an "Orbital Launch Facility" that would include fuel tankers, a basic crew habitat, and the possibility of later expansion into a full-fledged wheel-like space station.[44] A Sperry Rand Systems study prepared for Marshall left no doubt as to the facility's ultimate purpose: "The or-

DEPLOYMENT

Langley Research Center's Rene Berglund designed a space station using six rigid modules that were connected by inflatable passageways coming off a central nonrotating hub. The structure would self-deploy after being launched into orbit atop a Saturn V rocket. (NASA photo, no. 63-Space Station-1)

bital launch facility is the complex which is placed in orbit around the Earth. It can consist, in spectrum, of anything from a tool kit for OLV [orbital launch vehicle] assembly to a structure housing station keepers, crew members, and the equipment necessary for Apollo mission support, with gradual evolution to an advanced depot for lunar and planetary mission support."[45] This proposal represented an ingenuous way of getting both the lunar mission and a space station. What spaceflight advocates, including von Braun, had not foreseen was the implications of a race to the Moon. In that race time became the limiting resource, and such an orbital facility could not be completed in time to help with Apollo.

Even so, there were several reasons for which Wernher von Braun and his associates en-

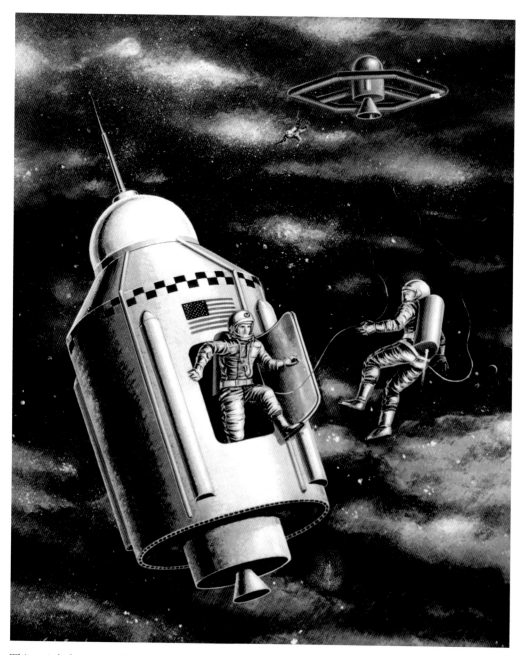

This artist's depiction of Berglund's self-deploying space station demonstrates the concept in orbit, complete with spacewalking astronauts in an Apollo-like capsule in the foreground. (NASA photo, no. 64-Space Station-24)

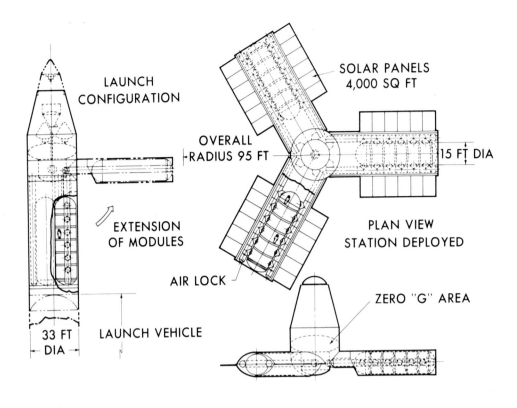

LAUNCH
CONFIGURATION

SOLAR PANELS
4,000 SQ FT

OVERALL
RADIUS 95 FT

15 FT DIA

EXTENSION
OF MODULES

PLAN VIEW
STATION DEPLOYED

AIR LOCK

ZERO "G" AREA

33 FT
DIA

LAUNCH VEHICLE

NASA-MSC E H OLLING 18 MAR 63 S-140-268

Another concept that gained currency in the mid-1960s was a self-deploying three-radial-module space station. Launched as a single collapsed structure in a Saturn V, it would deploy in orbit, begin spinning on a central axis, and house a crew of up to ten astronauts. (NASA photo, no. 63-Space Station-18)

thusiastically urged a decision in favor of Earth-orbit rendezvous: the direct ascent approach was technologically unfeasible before the end of the 1960s; their option provided a logical rationale for a space station; and it ensured an extension of the Marshall Space Flight Center's workload, a consideration that was always important to center directors competing inside the agency for personnel and other resources. At an all-day meeting on June 7, 1962, at Marshall, NASA leaders met to hash out these differences. The debate was heated at times, but after more than six hours of discussion, von Braun finally gave in to the lunar-orbit rendezvous mode, saying that its advocates had demonstrated adequately its feasibility and that any further contention would jeopardize the president's timetable.[46] Although this effectively settled the Apollo mode debate, von Braun and his associates continued long-range studies throughout the 1960s that they hoped would eventually lead to a space station.

2001: A SPACE ODYSSEY, HIGH POINT OF THE VON BRAUN SPACE STATION

The film *2001: A Space Odyssey*, released by director Stanley Kubrick in 1968, set a high mark for depicting a wheel-like space station on the von Braun model. Kubrick showed a human race moving outward into the solar system, and the station served fundamentally as its base camp. A great space station orbited the Earth, serviced by a reusable, winged spacecraft traveling from the globe's surface. Activities in low Earth orbit were routine in this world, with commercial firms carrying out many spaceflight functions. For example, the shuttle to the space station is flown by Pan American, a Hilton hotel is located on the station, and communications back to Earth are provided by AT&T.[47]

In this artist's conception of the three-radial-module space station in Earth orbit, Apollo hardware is used both to deploy the station and to ferry crews to and from orbit. (NASA photo, no. S-64-3704)

Kubrick's film was, from start to finish, a special-effects masterpiece, especially for the precomputer graphics era of filmmaking. Perhaps its most scintillating segment involved the docking of a winged shuttle with the gigantic rotating wheeled space station while the "Blue Danube" waltz provides the only sound. Kubrick's illustration of this station expressed well the von Braun dream. With its twin hubs still under construction, Kubrick's base camp to the stars measured an astounding 900 feet across as it spun in its orbit 200 miles above the Earth. It held an international crew of scientists, bureaucrats, and passengers on the way to and from the Moon.[48]

The impact of von Braun's station, especially as Kubrick presented it, has been nothing short of amazing. John Hodge, the leader of NASA's Space Station Task Force that worked to design a station in the 1980s, told Congress in 1983, "I think if you ask the public at large, and quite possibly most of the people within NASA what a space station was, they would think in terms of the movie that came out 15 or 20 years ago." He added that people expected a station to consist of "a very large rotating wheel with 100 people on it and artificial gravity."[49] Writing for the popular *Science 83* magazine, Mitchell Waldrop felt it necessary to explain why any real space station would not look like the von Braun version. He observed that the actual station "will look more like something a child would build with an Erector set." It would not resemble a wheel, it would not rotate, and it would not soar very high. "*2001* it's not," Waldrop explained.[50]

NASA engineers believed that the three-radial-module space station would be serviced by a scaled-up Apollo spacecraft that could carry six astronauts. (NASA photo, no. S-63-3536)

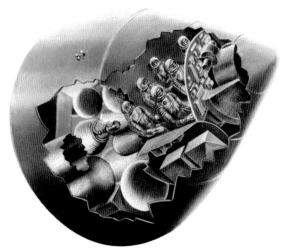

As the foregoing suggests, it would be difficult to overestimate the significance of the von Braun paradigm, especially its centerpiece of a space station as the base camp to the stars. Von Braun articulated a powerful image of the space station and sold it to his technological colleagues and the American public. He energized NASA to adopt the space station as a long-term goal, and even when other priorities have arisen, NASA has remained committed to it as necessary for space exploration. At every opportunity, the space agency has advocated for a space station as an orbital laboratory, observatory, industrial plant, launching platform, and drydock. Momentarily sidetracked by Project Apollo in the 1960s, space station advocates went underground. They pursued studies on how to use Apollo hardware in a space station program that would follow the Moon landings. The result became Skylab, a strikingly different type of station from that envisioned by Wernher von Braun, but a space station nonetheless. Its mission no longer explicitly stated as a jumping-off point for future space exploration, Skylab would serve as a preliminary station that would begin humanity's long tutorial on how to live and work in space.[51]

The quintessential space station in the minds of most Americans is the visually powerful image offered in *2001: A Space Odyssey*. This illustration shows a United States with permanent outposts in Earth orbit, a necessary way station to the Moon and planets. (NASA photo)

CHAPTER 3

Skylab and the Salyuts

Preliminary Space Stations

In the 1960s something remarkable happened. Both the United States and the Soviet Union abandoned the giant-wheeled space station emphasized in more than fifty years of studies and popularized by Wernher von Braun in favor of an entirely different type of structure. Each Cold War rival set about developing orbital workshops that would allow extended human missions in Earth orbit, but made no provision for artificial gravity through centrifugal force. Neither the American nor the Soviet efforts were intended as true space stations, in the sense that they might serve as base camps for exploration, but rather as orbital laboratories where humanity could harness one of the most unique attributes of spaceflight—microgravity—to assist in research. The American Skylab was launched in 1973. It was occupied in 1973 and 1974, and during three missions, three Apollo crews lived in the workshop for a total of 171 days and 13 hours.

In Skylab, both the hours in space and those spent in performance of extravehicular activity under microgravity conditions exceeded the combined totals of all the world's previous space-flights up to that time. NASA would have sent another crew to the Skylab workshop in 1978 had the Space Shuttle come on line early enough. Instead, in the fall of 1977, agency officials determined that Skylab had entered a rapidly decaying orbit—resulting from greater-than-predicted solar activity—and that it would reenter Earth's atmosphere within two years. They steered the orbital workshop as best they could so that debris would fall over oceans and unpopulated areas of the planet. On July 11, 1979, Skylab finally returned to the Earth's surface.[1]

At the same time, the U.S. Air Force sought a place for itself in Earth orbit and designed a military space station, the Manned Orbiting Laboratory (MOL). The Department of Defense (DoD), and in particular the Air Force, had been busy conducting its own space station studies since the late 1950s. There were experts in DoD who keenly saw the potential of military applications in Earth orbit. Although the project never reached construction phase, the MOL program represented an outriding effort from another federal agency to co-opt space for its purposes. This raised all manner of policy questions that had to be hashed out during the 1960s and early 1970s.[2]

The Soviet Union launched its first orbital workshop, Salyut 1, in 1971, and sent two cosmonaut crews there. Over the next eleven years, the Soviet Union launched a total of seven Salyut military space stations. The last of these, Salyut 7, was operational for almost three years when, on February 5, 1985, ground controllers lost all contact with it. In a spectacular mission, on June 6, 1985, the crew of Soyuz T-13 repaired the space station and continued its use until the deployment of Mir in 1986.[3]

This chapter tells the story of the Skylab and Salyut workshops, including the many design efforts that did not yield an orbital vehicle.

Defining a New Space Station Program

In the aftermath of the Apollo mode decision in 1962, everyone at NASA interested in a space station scrambled back to their centers to reconsider options. At no time did they abandon their belief in a station as a logical step in the process of human exploration. They modified configuration and purpose, in the former instance radically but in the latter modestly, and insisted that it remain the base camp to the stars. This dynamic has been repeated throughout human history, especially when belief systems crumble under scrutiny. Proponents rarely abandon their *Weltanschauung* when reality intrudes on it in discomforting ways. Instead, they alter their arguments to make them more compatible with the situation. They also seek out new beliefs to validate their old behavior and new explanations as to why initial concepts failed. Sometimes they even deny that their positions did, in fact, fail.[4]

At the end of July 1962, NASA's Langley Research Center hosted a NASA-only space station symposium to discuss how best to proceed in the aftermath of the Apollo mode decision. Center director Floyd L. Thompson later summarized the opinion of many in attendance when he wrote that the "space station program has a high degree of acceptability within NASA, Congress, etc. . . . [but the] Space Station is in competition with other programs in consideration of time and money, and from the Congressional and Administration viewpoint." Consequently, whatever NASA did in pursuing the effort, Thompson thought, should never receive undue funding and publicity until Apollo had reached a completion point.[5]

Thereafter, NASA continued to try to reach an internal consensus on how a space station of the future might satisfy a number of constituencies while remaining subliminal to the efforts of Apollo. But Thompson clearly understood that nothing would be allowed to circumvent Apollo, and any space station studies had to remain secondary to that larger effort. Joseph Shea, deputy director for systems of NASA's Office of Manned Spaceflight, held responsibility for defining an agency-wide long-range space station plan. He worked, unsuccessfully most of the time, with several NASA headquarters offices, each representing divergent areas of interest including science, space applications, and advanced research and technology. Shea sent out surveys in late 1962 to each of these offices, and in early 1963 he set up "use panels" with representatives from the offices to define the station plan. But the surveys received only token response, and some of the panels did not even meet. In frustration Shea realized his bureaucratic trap: without a formally approved and funded space station program, few within the agency wanted to spend their time and effort working on defining the effort. At the same time, without firm objectives and mission requirements, it seemed unlikely that NASA would ever obtain congressional authority to proceed with a program.[6] The result was that individual study efforts at several NASA centers went forward without any real direction.

The Era of Langley's MORL

Even though NASA officials chose lunar-orbit rendezvous as the mode for reaching the Moon in 1962, obviating the need for an Earth-orbital space station, diehards at Langley continued work on space stations, but their studies soon took an interesting and ultimately decisive turn away from von Braun's massive wheel. Emphasizing a smaller and more economical station that complemented and made maximum use of the technological systems being developed for Apollo, they came up

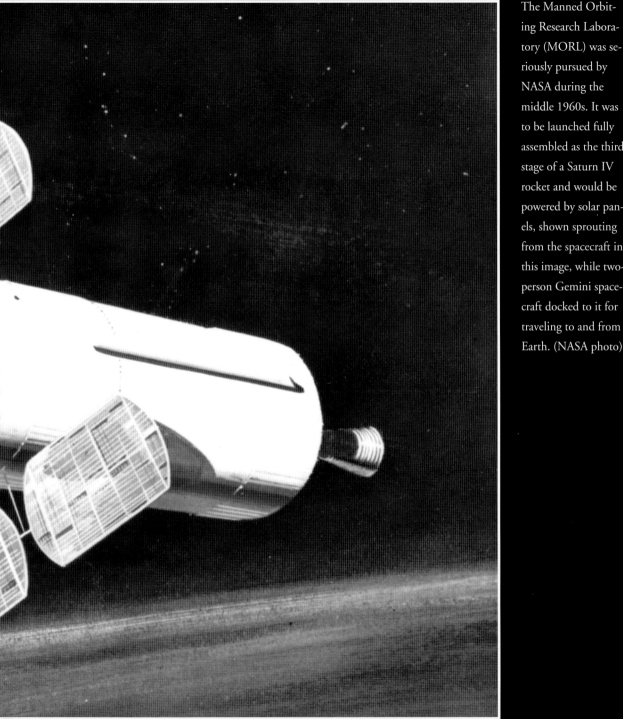

The Manned Orbiting Research Laboratory (MORL) was seriously pursued by NASA during the middle 1960s. It was to be launched fully assembled as the third stage of a Saturn IV rocket and would be powered by solar panels, shown sprouting from the spacecraft in this image, while two-person Gemini spacecraft docked to it for traveling to and from Earth. (NASA photo)

with the Manned Orbital Research Laboratory (MORL). In contrast to the gigantic space station so common in popular culture, this vehicle would not rotate and would house only a crew of up to nine astronauts. Its purpose, furthermore, emphasized cutting-edge research that exploited the microgravity environment of Earth orbit. Its designers did not necessarily view it as a base camp, although it might eventually, at some ill-defined future time, serve that purpose as well.

The original MORL concept called for a "minimum size laboratory to conduct a national experimental program of biomedical, scientific, and engineering experiments," with the laboratory specifically designed for launch atop a Saturn I or IB.[7] They established the MORL program's core goal as having one astronaut live in space for one year, with other crew members aboard for shorter periods on a rotating schedule. They envisioned this as attainable by 1965 or 1966 in unison with Project Gemini, the NASA program bridging Mercury and Apollo, whose basic purpose was to resolve the problems of rendezvous and docking and of the long-duration human spaceflights that would be necessary for a successful lunar landing.[8]

In this plan, a Saturn booster would launch MORL into Earth orbit. Then two astronauts in a Titan-launched Gemini spacecraft would "ascend to the laboratory's orbit and complete a rendezvous and docking maneuver." They would take up residence in the laboratory, and a few weeks later two more crew members would join them. One new astronaut would enter the laboratory at each crew change, thus providing a check on the cumulative effects of weightlessness. Three of the astronauts would occupy the space station for only parts of a year, but one of the original crew might complete a full year on orbit. To replenish consumables, every ninety days an Atlas-Agena combination would rendezvous with one of the laboratory's four docking ports.[9]

Despite the practicality of the effort, Langley could not obtain permission to move beyond these cursory studies. To reinstate MORL even on a provisional basis, Langley associate director Charles J. Donlan argued aggressively that a space station represented "the next logical step" after Apollo and that planning should begin immediately. Before interplanetary missions could be undertaken, he asserted, NASA had to learn much more about the long-term effects on human crews of weightlessness. An orbital laboratory offered the only good way of making such studies. Donlan's lobbying paid off. In 1963 NASA deputy administrator Robert C. Seamans Jr. authorized Langley to proceed with the first phase of the MORL design.[10]

Langley solicited Phase II design studies for proposed MORL designs.[11] Each potential contractor submitted work that adhered to the following standards:

1. Zero gravity (weightless conditions) should be the primary operating mode of the space station, but some capability for artificial gravity up to 1 g was "desirable."
2. Reliability should be "accomplished by minimum complexity" with "the maximum use of developed hardware," including "redundancy where needed."

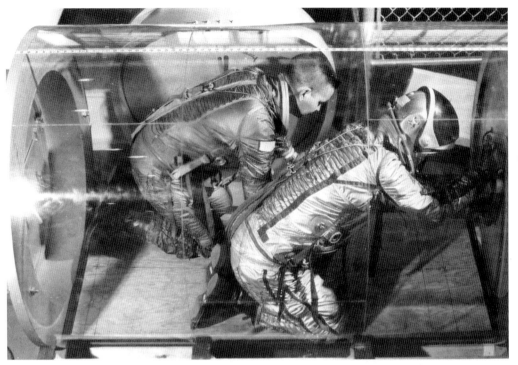

Future operations on a space station required NASA engineers to think seriously about airlocks and how to move between vehicles. This experiment in 1966 tested a Plexiglas space station airlock and whether humans could use it with only a pressure suit. (NASA photo, no. 66-H-428)

3. MORL should be launched on the Saturn I, Saturn IB, Titan II, or Atlas-Agena launch vehicles.

4. The lifetime of MORL should be a "minimum of one year manned occupancy."

5. The human crew should be rotated "at suitable intervals with one to stay a year."

6. The launch site should be Cape Canaveral, Florida, with "maximum use of existing and planned facilities."

7. A ferry spacecraft should always be available for evacuation of the entire crew.

8. MORL should also make "maximum use of launch, rendezvous, reentry, and landing procedures developed in existing programs."

9. MORL should use tracking, instrumentation, and data acquisition facilities being planned or under development.

10. Nothing about MORL should interfere with, interrupt, or delay the Apollo program.[12]

In essence, MORL would leverage existing systems as much as possible, limit new research and

The central question about the space station involved solving the problem of generating enough electrical power. Only two options emerged for accomplishing this task over the long term: solar power and nuclear power. The latter was enormously dangerous, and the former yielded only a small amount of wattage. But with additional research, both could be improved. In this illustration, a nuclear SNAP II reactor located at the end of a boom, away from the habitat, powers the station. The Marshall–McDonnell Douglas approach envisioned the use of two common modules as the core configuration of a small space station. Each would be 33 feet in diameter and 40 feet in length, and each would provide a building block not only for the space station but also for a much larger fifty-person space base. Coupled together, the two modules would form a four-deck facility: two decks for laboratories and two for operations and living quarters. Zero gravity would be the normal mode of operation, although the station would have an artificial-gravity capability. This general-purpose orbital facility would provide wide-ranging research capabilities. Its design was driven by the need to accommodate a broad spectrum of activities in support of astronomy, astrophysics, aerospace medicine, biology, materials processing, space physics, and space manufacturing. (NASA photo no. 63-Space Station-20)

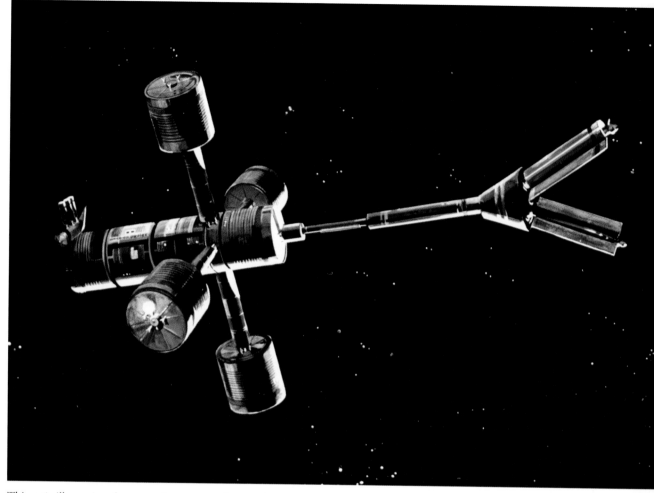

This 1969 illustration shows another concept for a nuclear-powered space station, again with the reactor located far away from the habitat. (NASA photo, no. MR69-6320)

development to those systems that did not already exist, maintain cost control, and avoid any effect on the lunar landing program.

Although most of these priorities reflected longstanding ideas expressed in earlier studies, MORL differed from them in its emphasis on developing a weightless facility. The American and Soviet experiences with human spaceflight had demonstrated the capability of astronauts to function in microgravity without deleterious effects, at least for several days. The laboratory would help determine how much longer one could extend human presence in zero gravity. Clearly, the thinking at Langley concluded that astronauts could stay in orbit for at least a year. And that was a belief based not on wishful thinking but on aerospace physicians' growing confidence that astronauts could do quite well in space at least for that long. The move to a weight-less facility recognized that astronauts would require proper conditioning on orbit, and the MORL program required Boeing Airplane Company and Douglas Aircraft Corporation to develop a centrifuge for intermittent astronaut use. This change, reflected for the first time in the 1963 MORL program decision, represented a critical shift in the American concept of a space station. It abandoned the classic theory, ensconced in the spaceflight gospel for more than half a century, that a space station must continually rotate to provide artificial gravity.[13]

In December 1963 NASA selected the engineering team at Douglas for the Phase II work on MORL, a nine-month study contract worth just over $1.4 million. Douglas engineers had shown a longstanding interest in space stations. In 1958 the company had received a $10,000 first prize in a contest for design of "A Home in Space" sponsored by the London *Daily Mail,* and in 1961 it had competed for the six-month study of Berglund's rotating hexagon concept.[14] The MORL baseline concept developed by Douglas proved a mixture of old and new ideas. Douglas's baseline facility could be carried into orbit as a unit aboard a Saturn launch vehicle. Powered by solar cells in an era before they had become efficient, it also sported a nuclear-power system. As Langley dictated, Douglas's MORL concept abandoned artificial gravity—and with it the need for a rotating structure—but it also consisted of a series of discrete modules. This modular approach made it possible to build MORL in stages, allowing it to grow and evolve with the capabilities of space technology and the needs of the space program.[15]

Further, MORL could serve the needs of space science. Douglas engineers, partly in response to Langley's attitudes, viewed MORL as a facility in which research could be carried out across the broad spectrum of scientific disciplines. In addition to serving as a platform for astronomical telescopes, the MORL could be designed to have a self-contained module for biological studies. Such research had potential applications not only in basic life sciences research but also in medicine and pharmaceuticals. For the aid of geologists, oceanographers, and meteorologists, Douglas provided a specialized nine-lens camera system for multiband spectral reconnaissance of Earth features and weather systems. It also allowed side-looking radar to garner data for large-scale topographical mapping. And it could prove the ideal place to study subsystems for interplanetary vehicles and their propulsion systems, technologies whose performance in space could not be simulated adequately on the ground. Douglas's proposal even envisioned using MORL as a platform in lunar orbit that could perform surface observation and mapping, landing site selection, and lunar module support. With the addition of a state-of-the-art landing stage, Douglas engineers insisted, MORL could even land on the Moon and provide a long-term base for exploration. Finally, MORL might even serve as the jumping-off point for a mission to Mars, and as a module of an orbital human planetary mission.[16]

Although many variations in size and weight of the basic MORL design emerged between

1963 and 1966, Langley never wavered in its goal of developing a single, highly integrated scientific laboratory to be placed in low Earth orbit that would last for at least one year. The MORL configuration as developed by Douglas Aircraft provided a "shirtsleeve" environment for the astronauts, living and working in four compartments mounted atop each other. At the top were the astronaut quarters, and occupying the compartment immediately below was a centrifuge for maintaining muscle in microgravity. The third level housed the control systems for the space vehicle, and the bottom contained the scientific laboratory.[17]

All of this sounded too good to be true, and it was. Although MORL had several positive features, it proved both too expensive to pursue and technologically too ambitious to receive much support at NASA while Apollo consumed the resources of the agency. Instead the MORL studies went into the technical library to be consulted by subsequent space station designers. But many of the ideas that first appeared in MORL did find their way into the Skylab program.

The Olympus Project and Other NASA Space Station Designs

As the MORL project proceeded along its design path in the mid-1960s, other space station concepts also received attention at NASA's Marshall Space Flight Center in Huntsville, Alabama, and the Manned Spacecraft Center (MSC) in Houston, Texas. All of these, furthermore, followed MORL's lead in abandoning artificial gravity as a maxim of the space station. The most ambitious was the Olympus project of the MSC.

In July 1962 two engineers at the MSC, Edward Olling and Maxime "Max" Faget, proposed building a "Large Orbiting Research Laboratory" (LORL) at a cost of $4 billion. Under this scheme, NASA would construct a space station "system" that could fly in 1967 using Saturn V boosters for the initial launch of the main station, Saturn IBs or Titan IIIs for resupply, and a modified reusable six-person Apollo logistics spacecraft called "Diana." The station was enormous, having a volume more than seven times that of MORL (67,300 cubic feet compared with only 9,000 cubic feet) and ten times the weight (74,600 pounds versus 6,800 pounds). Using this design, Olling and Faget believed, the station could house an astronaut crew of twenty-four over a useful life of five years.[18]

NASA leaders rejected the plan almost as soon as Olling and Faget proposed it. "The present state of knowledge on the uses for, and requirements of, a space station are so nebulous that to plan for a single type of sophisticated, large, many-manned space station is very shortsighted," they told the two MSC engineers. Faget assumed that the real reason Olympus received cursory attention was that it looked too expensive. He went back to the drawing board to reduce the cost, but he proved to be wrong about this problem. As NASA official Douglas Lord wrote, "They think if it's cheap enough, they may sneak it by." But it did not have sufficient justification regardless of the cost.[19]

In this 1964 artist's conception, a large wheeled station orbits high above Earth, powered by a Westinghouse electrical power plant. Electrically propelled space vehicles take astronauts to the Moon and planets. (Courtesy of Westinghouse, no. PR-25184)

At Marshall, von Braun's team pursued development of a space station that also played off of the MORL concept to some degree. Aerospace engineer Heinz Herman Koelle and his associates devised a scenario in which a spent upper stage of the Saturn V could be modified in Earth orbit for permanent habitation. Koelle advocated what von Braun characterized as a "poor man's approach" to building a space station.[20] It steered a path away from large concepts such as Olympus and MORL as far as possible, while trying to remain under an intangible cost level where it would not excite attention from NASA leaders. Ultimately it failed in that objective, but by then NASA was ready to pursue Skylab, and this concept became the core of design work that went into that orbital workshop.

In the end, although details of these studies may have proven fascinating to engineers working on them, they all pointed in the same direction. All showed that the giant wheel-like space station of the 1950s could neither be built for a reasonable price nor was it necessary. The studies also demonstrated two distinct approaches to constructing a space station. The first involved building a large vehicle that could cost as much as $4 billion in 1964 dollars. This structure, exemplified by MORL and Olympus, would lead to a multipurpose facility capable of producing significant scientific and technological data. A second approach suggested that slightly modifying Gemini and Apollo program hardware for extended space operations represented little risk and expense—only about $1 billion—but with limited objectives and results.[21]

With this in mind, NASA personnel set about designing a tiny space station known as Apollo X, really just an Apollo spacecraft modified with additional life support systems for an extended orbital mission. This space station had only 600 cubic feet of volume and was based entirely on Apollo technology. Apollo X would be a small "limited life" laboratory, serving a crew of two over a useful life of only 30 to 120 days. In between the Olympus and Apollo X rested an idea for an AORL, or Apollo Orbital Research Laboratory: a medium-size station (5,600 cubic feet in volume) that could enjoy an extended life of two years, with a crew from three to six astronauts.[22]

All of these efforts applied a modular concept to the space station, following the lead of Darrell Romick and Heinz Koelle in the 1950s. Engineers proposing this approach frequently used the analogy that building a space station was similar to the manner in which children assembled Tinker Toys to make complex structures. Each module would be sent up individually and assembled on orbit, but the main argument against this was that space assembly had not yet been tested. In 1963 the piloted or automatic rendezvous of two spacecraft and the manual assembly outside a spacecraft of prefabricated parts were unknowns.[23] But the modular concept appealed to managers and engineers alike, because it allowed the use of Apollo technology, saved resources, and provided flexibility in the face of uncertain program objectives and an even more uncertain budget. NASA returned to the modular approach several times and eventually used it for the International Space Station.[24]

A concept for the proposed U.S. Air Force's Manned Orbiting Laboratory in the mid-1960s. (NASA photo)

The Military Space Station

Since the early 1950s, some quarters of the U.S. Air Force had expressed interest in placing military personnel in orbit.[25] With the development of robotic satellite technology in the early 1960s, the need for humans on military space vehicles diminished—but not the desire of those in the Air Force who wanted "a piece of the NASA action." The result was an Air Force program to build a military space station, the Manned Orbiting Laboratory. Langley engineers quickly realized that MOL was essentially MORL with the research left out. The space vehicle could be used for biomedical experiments on the crew in a weightless state, but for little else.

In October 1963 Secretary of Defense Robert S. McNamara announced DoD support for an Air Force plan to build MOL, never mind the successful efforts of the United Nations to broker a deal between the United States and the Soviet Union that forbade weapons of mass destruction in outer space. Testing the military usefulness of humans in orbit, McNamara argued, could only be done through the development of a MOL vehicle based on a combination of Gemini and Titan IIIC hardware. It would be inexpensive, he commented, and provide important scientific data on the ability of humans to live in space.[26]

The decision to proceed with MOL represented a direct incursion into one of NASA's core activities—human spaceflight—especially because of MOL's intended role of gathering biomedical data on weightlessness. NASA administrator James E. Webb immediately sought out McNamara to discuss roles and missions. In a joint document, DoD and NASA agreed that the Air Force MOL program would "not supplant the possible longer term requirement for an orbital space station which would be capable of meeting NASA's and DoD's future respective needs."[27] But this compromise raised some important questions. If MOL was not a "true" space station, then what was? What defined a national space station program that set it apart from the kind of program that the Air Force might develop?

Despite these questions, the Air Force moved ahead on the MOL design effort. The baseline configuration used a two-person Gemini B spacecraft attached to a laboratory vehicle that could be launched as a single unit atop a Titan IIIC. The spacecraft would be launched from Cape Canaveral into a 125 to 250 nautical-mile equatorial orbit inclined less than 36° to the equator. This ensured that the orbit did not pass over any part of the Soviet Union and that MOL could not be used for reconnaissance, something the Air Force took care to ensure that the Soviet Union understood. Once in orbit, the crew would transfer to the laboratory vehicle through a hatch cut in the Gemini B's heatshield. The crew could then spend up to one month in space. At the end of the mission the crew would transfer back to the Gemini B, reactivate its systems, reenter the atmosphere, and splash down in the ocean.[28]

The MOL development program was a disaster from the beginning. It endured technological complications, schedule delays, cost overruns, and reliability difficulties. To get out of the morass that had become MOL by 1966, some Air Force officers suggested partnering with NASA on the program. After repeated failures, the service finally successfully tested a Gemini heatshield on November 3, 1966. But it was a minor balm, and the program continued to creak forward with all manner of problems. As things dragged on, the operational date kept slipping. Although it was to be flown by then, in late 1966 the Air Force announced that the first of five piloted MOL launches would not begin before December 15, 1969. But as that year approached, the service pushed back the first MOL flight to fall 1971.

By 1969 the original cost of the program—estimated at $1.5 billion—had risen to more than $3 billion. At the same time, any legitimate mission for "blue-suiters" in space had diminished. Finally, on June 10, 1969, Defense Secretary Melvin R. Laird informed Congress that MOL would be canceled, saying that the DoD needed to reduce spending. He also noted the advances in performance and reliability of instrumented military satellites, obviating the necessity of humans.

By its end, the MOL program had sucked up $1.3 billion that might have been productively spent on any of the NASA space station concepts. About the only worthwhile result of MOL was that several of the military astronauts selected for the program transferred to NASA and became the first to fly the Space Shuttle. Among them was Richard R. Truly, a naval aviator. He flew shuttle approach and landing tests in the 1970s, flew several orbital missions in the early 1980s, led the NASA effort to recover from the Challenger accident, and went on to serve as the agency's administrator between 1989 and 1992.[29]

APOLLO APPLICATIONS PROGRAM:
THE ORIGINS OF SKYLAB

Capitalizing on the creation in August 1965 of the Saturn-Apollo Applications Program Office at NASA headquarters—established to promote use of Apollo hardware for future projects—advocates of a space station finally seized the initiative, and the budget, for a new project that would take up once again the effort to create a permanent human presence in Earth orbit.[30] The project became Skylab, and for two years in the early 1970s it served as the centerpiece of NASA's effort to explore the cosmos. It began with moderate hopes, reaped a rich scientific harvest, and died a martyr's death by fiery reentry in 1979. Skylab was never the project that NASA intended as a space station, but it filled an important gap between what was possible and what was desired.

Skylab emerged in the middle 1960s out of the Apollo Extensions Support Study that sought to find new or modified flight projects using the spacecraft and launch vehicle technology developed for the lunar landing. Practical use of Apollo hardware, therefore, became the modus operandi for virtually every follow-on program that NASA considered during the era. Skylab was no exception. One of the possibilities considered from the first was an Apollo Command and Service Module to carry an assembly of small solar telescopes into orbit, there to be deployed and operated on the service module with the assistance of the astronauts. The Apollo spacecraft would then

Another space station concept that the Department of Defense championed during the 1960s was Blue Gemini, depicted here in artist renderings showing a cutaway of the habitation module. (NASA photo)

return the exposed film to Earth. In 1963 this assembly was named Apollo Telescope Mount, a strange contraption that deployed from the service module and could be pointed toward various targets in the solar system and beyond. It soon morphed into a more aggressive suite of missions that included long-duration Earth orbital flights during which astronauts would carry out scientific, technological, and engineering experiments. After completion of the Apollo program, spacecraft and Saturn launch vehicles originally developed for lunar exploration would be further modified to provide the capability for crews to remain in orbit for extended periods.[31]

The Apollo Telescope Mount was the most important single scientific device on the orbital workshop. ATM was much more than a telescope mount. It was a full-size solar observatory with sophisticated technology continuously pointed at the Sun. Scientist-astronauts on board Skylab selected the targets on the Sun and controlled the instruments. Some used film, thereby improving spatial and spectral resolution. The crew monitored active regions of the Sun; when they saw signs of flare they focused the instruments on the region, turned on the cameras, and obtained a continuous record of the radiation emitted during a flare. In addition to catching solar flares in action, the astronauts completed several repairs during the flight, enabling all the instruments to work throughout the six-month experimental life of Skylab in 1973–74.[32]

Perhaps the most important decision in the early history of the Skylab program came on August 20, 1965, when Wernher von Braun gained formal approval for the Marshall Space Flight Center to proceed with serious investigation of the conversion of a spent Saturn S-IVB stage into a workshop housing an Apollo crew. This became known as the "wet" concept, because astronauts in space would refurbish for habitation a Saturn upper stage that had already been used to achieve orbit. The empty hydrogen tank would be purged and filled with a life-supporting atmosphere on orbit. Marshall Space Flight Center (MSFC) project leaders then invited others to help with a short-term study on the feasibility of the concept. Working with engineers at the Manned Spacecraft Center (which in 1973 would be renamed the Johnson Space Center) and Douglas Aircraft Company, builder of the upper stage, the three teams reached independent assessments. They gathered at Marshall on October 20, 1965, to present their findings, which agreed that the wet concept had excellent potential.[33]

Central to the objectives of the program was the use of Apollo technology, the equipment that NASA planned to be flying for more than a decade after the lunar landings had been completed. This would, NASA leaders argued, leverage the investment in Apollo hardware far beyond the Moon missions and open space to practical human applications. Then there was the science: physical, astronomical, and biomedical.

Some scientists wanted to use zero-gravity conditions for materials research. Understanding the manner in which materials react to the vacuum, temperatures, and physical properties found in space enticed some to embrace this program as a great boon to scientific understanding. More important, some viewed the orbital workshop as a harbinger of great commercial applications that would eventually make use of the microgravity found in Earth orbit, such as for crystal growth.

The ability to make more exact observations of both Earth and the solar system than ever possible beforehand also spurred interest in the orbital workshop and imposed some requirements on its design. Solar scientists had long dreamed of observing the Sun without interference from Earth's atmosphere. The shell of atmospheric gases that nourish and protect Earth severely limits an ability to view celestial objects, even with the finest and largest telescopes. An observatory orbiting above the Earth's atmosphere and equipped with fine resolution instruments, many believed, would be an extremely valuable tool for solar scientists. As studies for the orbital workshop progressed, plans also developed for more elaborate observations of the Sun with a group of solar telescopes mounted on Apollo-related spacecraft.[34]

But the most significant scientific windfall possible from the orbital workshop, in the minds of NASA officials, involved learning how to live and work in space. In 1965 the Gemini VII flight crew of Frank Borman and James Lovell completed a two-week mission, and Russian cosmonauts orbited Earth for eighteen days in Soyuz 9. From those missions, the longest anyone had yet undertaken, it was clear that bioscientists had much to learn about the ability of astronauts to adapt readily to spaceflight. The lengthy operation of the orbital workshop, where crews would spend as much as eight weeks in space, would go far toward filling in this lack of knowledge. At that time, long-range physiological and psychological effects of weightlessness on humans were far from fully understood, and some questioned whether even highly trained and superbly conditioned astronauts could perform the varied tasks expected of them.[35]

As they eventually coalesced, the objectives of the Skylab orbital workshop program included the following:

> Skylab missions have several distinct goals: conduct of earth resources observations; advance scientific knowledge of the sun and stars; study the effects of weightlessness on living organisms, particularly man; study and understand methods for the processing of materials in the absence of gravity. The Skylab mission utilizes man as an engineer and as a research scientist, and provides an opportunity for assessing his potential capabilities for future space missions.[36]

Carried out on a shoestring, for a human spaceflight program, made possible in large measure because of the use of equipment developed and built with Project Apollo funding, the direct Skylab expenditure was $2.5 billion.

By March 1966, NASA engineers envisioned three experimental modules consisting of Saturn S-IVB spent stages to be converted to workshops, and four Apollo Telescope Mounts.

They still believed that the wet concept was the best way to handle the challenge at that time. According to plans then being developed, a Saturn IB launch vehicle would carry an Apollo spacecraft into low Earth orbit. After the S-IVB stage had used up its fuel, astronauts in the Apollo capsule would dock and enter the stage's hydrogen tank through an airlock. They would then rid the tank of excess hydrogen, a not-inconsiderable task in itself, and refill it with oxygen. The tank would provide a bare workspace for the crew, without formal quarters and only the simplest of environmental controls. This first crew would undertake familiarization activities and work to learn how to move about in a controlled and enclosed microgravity atmosphere.

This concept remained the norm until a May 21, 1969, planning meeting held at the Manned Spacecraft Center in Houston. Engineers proposed constructing the workshop on the ground,

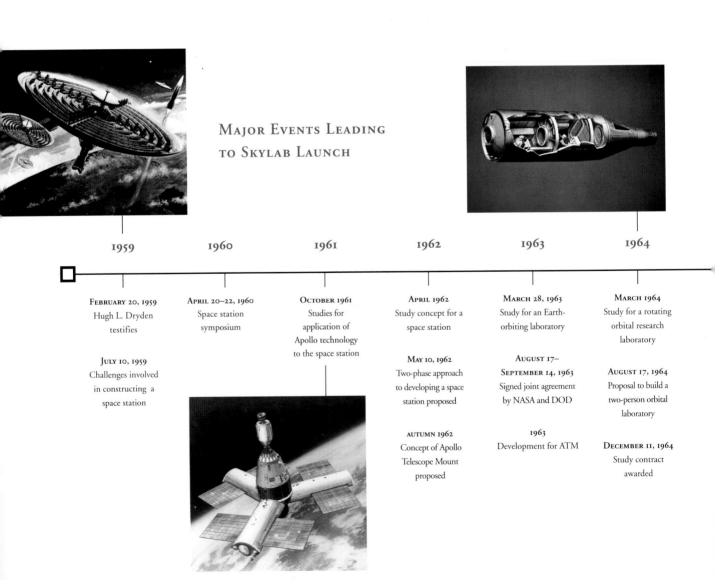

Major Events Leading to Skylab Launch

1959	1960	1961	1962	1963	1964

February 20, 1959
Hugh L. Dryden testifies

July 10, 1959
Challenges involved in constructing a space station

April 20–22, 1960
Space station symposium

October 1961
Studies for application of Apollo technology to the space station

April 1962
Study concept for a space station

May 10, 1962
Two-phase approach to developing a space station proposed

Autumn 1962
Concept of Apollo Telescope Mount proposed

March 28, 1963
Study for an Earth-orbiting laboratory

August 17–September 14, 1963
Signed joint agreement by NASA and DOD

1963
Development for ATM

March 1964
Study for a rotating orbital research laboratory

August 17, 1964
Proposal to build a two-person orbital laboratory

December 11, 1964
Study contract awarded

to the same external specifications as the S-IVB stage, and launching it atop the mighty Saturn V Moon rocket. The Saturn V had rather unexpectedly become available because the NASA administrator, to save money for the agency, had canceled the last three lunar landings. Debate followed, studies evolved, and on July 18, 1969, NASA administrator Thomas O. Paine approved a plan to build the workshop as a "dry" concept, assembled on the ground, and launch it intact.[37]

Designing Skylab

To reach that point of decision, as debate took place over the wet-versus-dry workshop concepts, design progressed. At a meeting at the Marshall Space Flight Center on August 19, 1966, for example, NASA associate administrator for Manned Space Flight George E. Mueller established

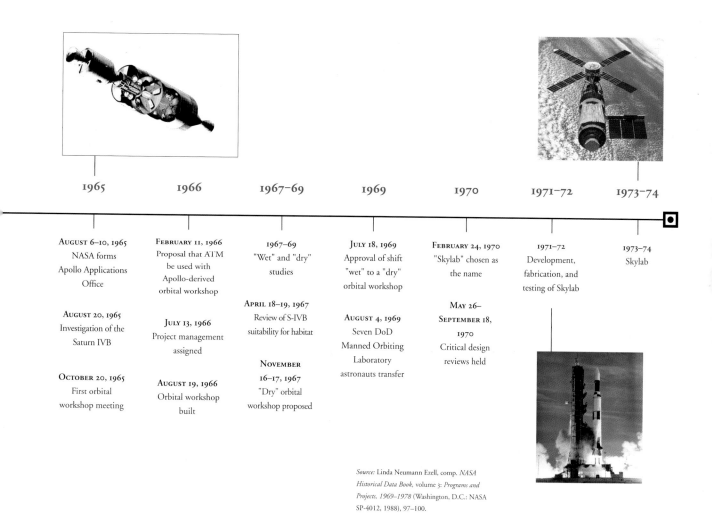

1965   1966   1967–69   1969   1970   1971–72   1973–74

August 6–10, 1965
NASA forms
Apollo Applications
Office

August 20, 1965
Investigation of the
Saturn IVB

October 20, 1965
First orbital
workshop meeting

February 11, 1966
Proposal that ATM
be used with
Apollo-derived
orbital workshop

July 13, 1966
Project management
assigned

August 19, 1966
Orbital workshop
built

1967–69
"Wet" and "dry"
studies

April 18–19, 1967
Review of S-IVB
suitability for habitat

**November
16–17, 1967**
"Dry" orbital
workshop proposed

July 18, 1969
Approval of shift
"wet" to a "dry"
orbital workshop

August 4, 1969
Seven DoD
Manned Orbiting
Laboratory
astronauts transfer

February 24, 1970
"Skylab" chosen as
the name

**May 26–
September 18,
1970**
Critical design
reviews held

1971–72
Development,
fabrication, and
testing of Skylab

1973–74
Skylab

Source: Linda Neumann Ezell, comp. *NASA Historical Data Book,* volume 3: *Programs and Projects, 1969–1978* (Washington, D.C.: NASA SP-4012, 1988), 97–100.

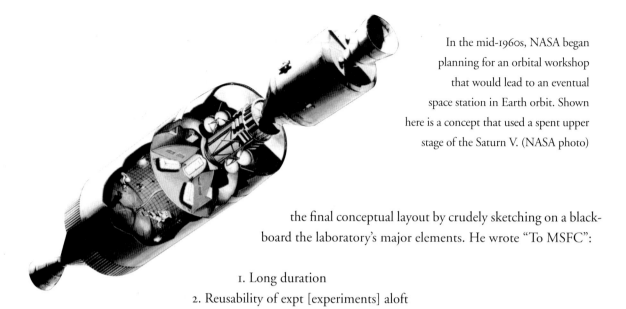

In the mid-1960s, NASA began planning for an orbital workshop that would lead to an eventual space station in Earth orbit. Shown here is a concept that used a spent upper stage of the Saturn V. (NASA photo)

the final conceptual layout by crudely sketching on a blackboard the laboratory's major elements. He wrote "To MSFC":

1. Long duration
2. Reusability of expt [experiments] aloft
3. Logistic resupply[38]

These priorities drove the design requirements thereafter.

What became the Skylab cluster was the largest spacecraft placed in Earth orbit up to that point. With the Apollo spacecraft docked to it, the system was approximately 117 feet long and had a mass of 199,750 pounds. It also contained a habitable volume of about 12,700 cubic feet. The major elements of the orbital workshop included the following five pieces:

Orbital Workshop (OWS) housing the crew, most of the stored expendables, a large experiment area, large solar arrays, and cold gas tanks and thrusters for secondary attitude control

Airlock Module (AM) with an airlock for extravehicular activities, main systems for communication and data transmittal, environmental and thermal control systems, and electrical power control systems

Multiple Docking Adapter (MDA) to provide docking ports for the Apollo spacecraft, housing also the control console for the Apollo Telescope Mount, controls and sensors for Earth resources viewing, and a number of other experimental facilities

Apollo Telescope Mount (ATM) carrying solar telescopes, control moment gyros (CMG) for primary attitude control, and four solar arrays for power

Instrument Unit (IU) holding necessary navigation, control, communications, and atmosphere systems

The most important element of this vehicle, though all proved integral, was the orbital work-

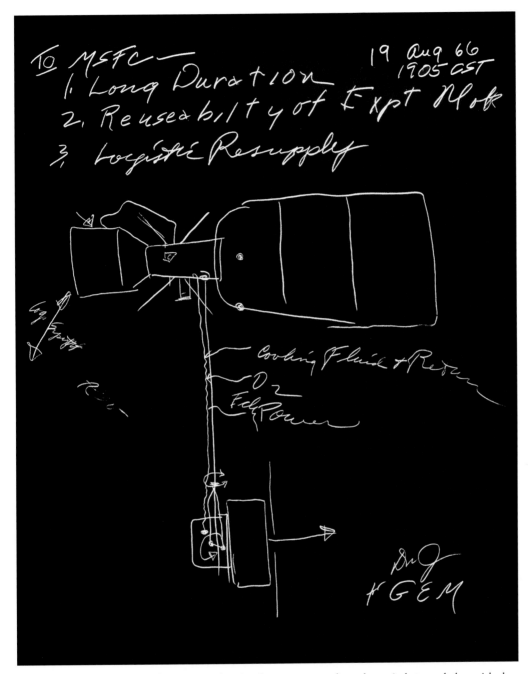

Seldom in aerospace history has a major decision been as promptly and concisely recorded as with the Skylab shown in this sketch. At a meeting at the Marshall Space Flight Center on August 19, 1966, George E. Mueller, NASA associate administrator for Manned Space Flight, laid out the final conceptual layout for the budding station's major elements. This was established as the Skylab in 1970. (NASA photo, no. MSFC-8886938)

shop. It consisted of two compartments or "floors," although such a term is largely meaning-less in a microgravity environment, separated by a perforated wall. The rear compartment contained a "wardroom" (named by astronauts with naval experience) with a section for food preparation and eating, a sleep section, a toilet and hygiene facility, and to one side a biomedical laboratory. The forward compartment was devoted primarily to experiments requiring relatively large volumes, or designed to use one of the two scientific airlocks for external viewing or exposure. It also included storage facilities; containers for food, water, and clothing; and a number of subsystems. Because of the weightless conditions, the workshop had handles, grips, and foot restraints, all mounted at appropriate places throughout the two compartments to enable the crew to move and secure themselves at workstations. Attached to the outside wall were two large, wing-like solar arrays.[39]

A final element of the assemblage, an Apollo spacecraft docked to the workshop, provided attitude, guidance, and control for the entire orbital laboratory. NASA included no thrusters on the workshop itself, a decision that no one realized at the time would prove quite significant when Skylab reentered the atmosphere in 1979. The Apollo spacecraft also provided the lifeboat that every vessel requires, and in the event of an emergency the crew could "abandon ship" and return to Earth aboard the self-contained capsule. A modified heatshield, coated with somewhat less ablative material than the lunar missions had required, protected the capsule against the blaze of reentry.

The program officially became Skylab on February 24, 1970. The contraction of "laboratory in the sky" had been suggested by Lieutenant Colonel Donald L. Steelman, who served at NASA on detached duty from the U.S. Air Force. He proposed the name in mid-1968, but NASA leaders decided to postpone renaming the program until funding seemed secure. Steelman later referred it to the NASA Project Designation Committee, which approved the formal name change on February 17, 1970. It proved a happy choice, for the name caught the imagination of the world.[40]

Most of the design and manufacturing for Skylab took place under contract to McDonnell Douglas Aircraft Corporation, Martin Marietta, and North American Rockwell, all giants of the aerospace industry. From the beginning of the project, an extremely close cooperation existed between NASA, scientific investigators, and industry, extending over all levels of the program structure. So highly integrated were the work teams that it often would have been difficult to differentiate between contractors, NASA engineers, and university scientists as the project proceeded from its early concepts through development, manufacturing, testing, final assembly, and checkout.

Nixon and the Quest for a "Real" Space Station

Even as Skylab was being built, NASA leaders recognized that it could not serve as the base camp that they had long wanted. Accordingly, in February 1969, when President Richard M. Nixon asked Vice President Spiro T. Agnew to chair a Space Task Group to provide "definitive recommendation on the direction which the U.S. space program should take in the post-Apollo period," NASA officials courted the group for a recommendation to build a full-fledged space station.[41] NASA administrator Thomas O. Paine tried to get an early commitment to what NASA saw as its major post-Apollo program. The need for an orbital outpost had been part of NASA's planning from the earliest years, but the decision to go to the Moon, particularly using a lunar rendezvous approach, had bypassed this step in space development. As NASA began during 1967 and 1968 to focus attention on its priorities for the next large new program after Apollo, a space station rose to the top of its list. On February 26, Paine sent to the president a lengthy memorandum on "Problems and Opportunities in Manned Space Flight." He suggested that "the case that a space station should be a major future U.S. goal is now strong enough to justify at least a general statement on your part."[42]

Paine found in Agnew an ally willing to advocate on behalf of a station, viewing it as a necessary prerequisite for lunar colonization and a mission to Mars. But many others within the Nixon administration failed to see not only the value of grand exploration plans but also the need for building an infrastructure such as a space station to undertake them. Most of these misgivings revolved around budgetary priorities, but some addressed larger questions. Former NASA deputy administrator and current secretary of the Air Force Robert C. Seamans, for example, recommended that NASA's program for the 1970s concentrate on using its capabilities for "solution of the problems directly affecting men here on earth," rather than for the development of a space station or anything similar.[43]

Agnew's Space Task Group, however, adopted an expansive vision and presented several future program options to President Nixon. Its report, delivered in September 1969, outlined three possibilities, each incorporating a space station and a Mars mission, but on different schedules and budget profiles. The report largely endorsed the development of a station and a reusable transportation system, emphasizing a "basic goal of a balanced manned and unmanned space program conducted for the benefit of all mankind." It did not abandon "the long-range option or goal of manned planetary exploration with a manned Mars mission before the end of this century as the first target," but deferred the necessity of the president approving an aggressive space station development schedule. In so doing, it left to President Nixon the job of setting the future course in space for the United States.[44]

Although the recommendation for a space station as a base camp remained a centerpiece of the report, Nixon chose not to accept that advice. Instead, he was silent on the future of the U.S. space program for more than a year after entering the Oval Office. Finally, Nixon issued a March 7, 1970, statement that clearly announced his approach toward dealing with NASA and its plans for aggressive space exploration: "[W]e must also recognize that many critical problems here on this

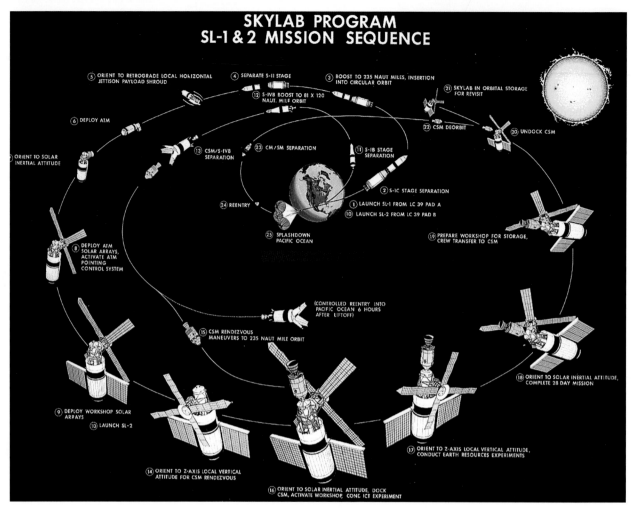

Seen here is the Skylab 1 and Skylab 2 mission sequence. The goals of the project were to enrich our scientific knowledge of the Earth, Sun, stars, and cosmic space; to study the effects of weightlessness on living organisms, including humans; to study the effects of the processing and manufacturing of materials utilizing the absence of gravity; and to conduct Earth resource observations. (NASA photo, no. MSFC-0100728)

planet make high priority demands on our attention and our resources."[45] The only item in the report that eventually received Nixon's endorsement was the Space Shuttle, the winged, reusable vehicle envisioned as a ferry between Earth and a space station.

Launching Skylab

The 100-ton Skylab 1 orbital workshop lifted off from NASA launch facility Kennedy Space Center, Florida, on May 14, 1973. (Kennedy the facility was thus named in November 1963, when

SKYLAB ORBITAL WORKSHOP

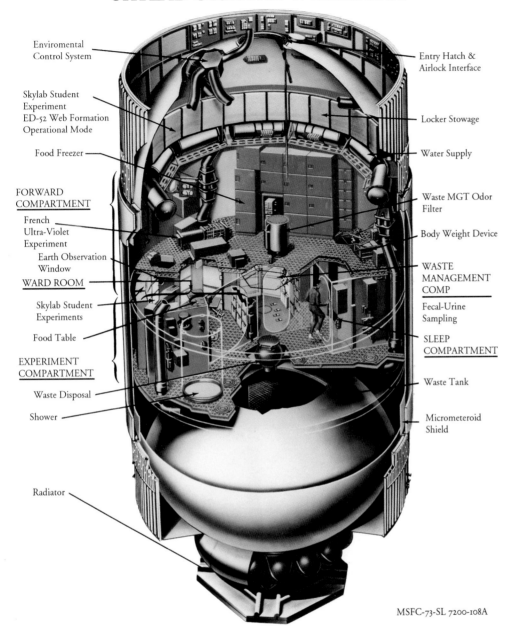

Enviromental
Control System

Skylab Student
Experiment
ED-52 Web Formation
Operational Mode

Food Freezer

**FORWARD
COMPARTMENT**

French
Ultra-Violet
Experiment

Earth Observation
Window

WARD ROOM

Skylab Student
Experiments

Food Table

**EXPERIMENT
COMPARTMENT**

Waste Disposal

Shower

Radiator

Entry Hatch &
Airlock Interface

Locker Stowage

Water Supply

Waste MGT Odor
Filter

Body Weight Device

**WASTE
MANAGEMENT
COMP**

Fecal-Urine
Sampling

**SLEEP
COMPARTMENT**

Waste Tank

Micrometeroid
Shield

MSFC-73-SL 7200-108A

This cutaway view of the Skylab Orbital Workshop (OWS) shows details of the living and working quarters. The OWS was divided into two major compartments. The lower level provided crew accommodations for sleeping, food preparation and consumption, hygiene, waste processing and disposal, and performance of certain experiments. The upper level consisted of a large work area and housed water storage tanks, a food freezer, storage vaults for film, scientific airlocks, and experimental equipment for mobility and stability and other research. The compartment below the crew quarters was a container for liquid and solid waste and trash that accumulated throughout the mission. A solar array, consisting of two wings covered on one side with solar cells, was mounted outside the workshop. This generated electrical power to augment that generated by another solar array, mounted on the solar observatory. Thrusters were provided at one end of the workshop for short-term control of the attitude of the space station. (NASA photo, no. MSFC-0101590)

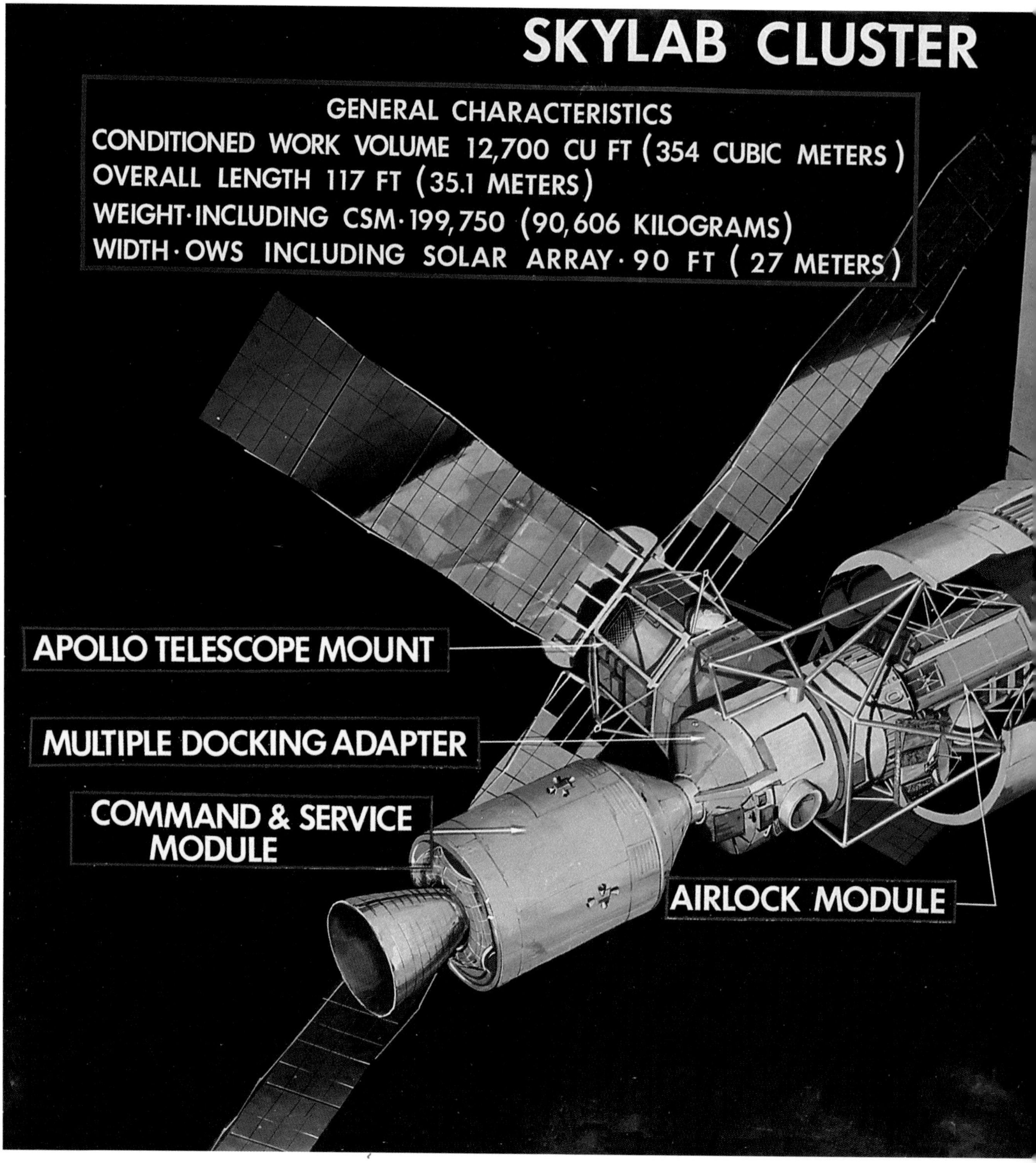

SKYLAB CLUSTER

GENERAL CHARACTERISTICS
CONDITIONED WORK VOLUME 12,700 CU FT (354 CUBIC METERS)
OVERALL LENGTH 117 FT (35.1 METERS)
WEIGHT·INCLUDING CSM·199,750 (90,606 KILOGRAMS)
WIDTH·OWS INCLUDING SOLAR ARRAY·90 FT (27 METERS)

APOLLO TELESCOPE MOUNT

MULTIPLE DOCKING ADAPTER

COMMAND & SERVICE
MODULE

AIRLOCK MODULE

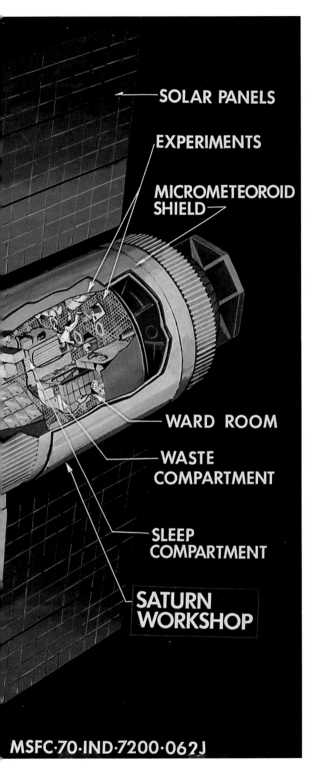

SOLAR PANELS

EXPERIMENTS

MICROMETEOROID SHIELD

WARD ROOM

WASTE COMPARTMENT

SLEEP COMPARTMENT

SATURN WORKSHOP

MSFC·70·IND·7200·062J

This illustration shows general characteristics of the Skylab, with callouts of its major components. In an early effort to extend the use of Apollo for further applications, NASA established the Apollo Applications Program (AAP) in August 1965. (NASA photo, no. MSFC-0101537)

Congress also redesignated Cape Canaveral, the geographic place, as Cape Kennedy. The facility's name has not changed since, but the place was later returned to its former name.) This was the last use of the giant Saturn V launch vehicle, which would place Skylab in orbit where a crew of three astronauts would later dock with it. Almost immediately, technical problems developed due to vibrations during liftoff. Sixty-three seconds after launch, the meteoroid shield—designed also to shade Skylab's workshop from the Sun's rays—ripped off. The accident investigation team reported:

> This was followed by the loss of one of the two solar array systems on the workshop and a failure of the interstage adapter to separate from the S-II stage of the Saturn V launch vehicle. The investigation reported herein identified the most probable cause of this flight anomaly to be the breakup and loss of the meteoroid shield due to aerodynamic loads that were not accounted for in its design. The breakup of the meteoroid shield, in turn, broke the tie downs that secured one of the solar array systems to the workshop. Complete loss of this solar array system occurred at 593 seconds when the exhaust plume of the S-II stage retro-rockets impacted the partially deployed solar array system. Falling debris from the meteoroid shield also damaged the S-II interstage adapter ordnance system in such a manner as to preclude separation.[46]

In spite of this, the space station achieved a near-circular orbit at the desired altitude of 270 miles.

To correct the problem, NASA engineers took two parallel paths. First, flight controllers maneuvered Skylab so that solar panels on the Apollo Telescope Mount faced the Sun to provide as much electricity as possible. But with the loss of the meteoroid and thermal shields, this allowed temperatures inside to rise to 126° Fahrenheit.

Second, engineers at the (by now renamed) Johnson Space Center drew up plans for a foldable screen to be deployed by the first crew that should shade the laboratory from the Sun's rays sufficiently to allow habitation. NASA even asked the National Reconnaissance Organization, then a classified entity, to "task" a reconnaissance satellite to photograph the Skylab space sta-

The launch of SA-513, a modified two-stage Saturn V vehicle, placed the Skylab cluster into Earth orbit on May 14, 1973. Skylab consisted of an orbital workshop, an Airlock Module, a Multiple Docking Adapter, an Apollo Telescope Mount, and an Instrument Unit. (NASA photo, no. MSFC-7040208)

Sixty-three seconds into the launch of Skylab, engineers in the operation's support and control center saw an unexpected telemetry indication that signaled damages to one solar array and the micrometeoroid shield of the orbital workshop. The micrometeoroid shield, a thin cylinder surrounding the workshop that protected it from tiny space particles and the sun's scorching heat, had ripped loose from its position. This had caused the loss of one solar wing and had jammed the other. Still unoccupied, the Skylab was stricken with the loss of the heat shield, and sunlight beat mercilessly on the lab's sensitive skin. Internal temperatures soared, rendering the station uninhabitable, threatening foods, medicines, films, and experiments. This image, taken during a fly-around inspection by the Skylab 2 crew, shows the exterior skin of the workshop discolored by solar radiation. (NASA photo, no. MSFC-0100554)

tion soon after launch, in order to assess the damage. These photographs were used to train the NASA astronauts who flew the repair mission.[47] Also responding to the crisis, the G. T. Schjeldahl Company of Northfield, Minnesota, manufactured a parasol made of Mylar, nylon, and aluminum foil and rushed it to the Kennedy Space Center for placement over Skylab by the astronauts during extravehicular activity (EVA). NASA had to postpone launch of Skylab 2, carrying the first crew to the station, so that engineers could develop hardware and crew procedures to make the workshop habitable. At the same time, NASA rolled and repositioned Skylab to equalize overall temperatures.

In an intensive ten-day period, NASA developed strategies and trained the crew. Finally, on May 25, 1973, astronauts Charles Conrad Jr., Paul J. Weitz, and Joseph P. Kerwin lifted off from Kennedy Space Center in an Apollo capsule atop a Saturn IB and rendezvoused with the orbital workshop. This Skylab 2 mission carried a parasol, tools, and replacement film to repair the orbital workshop. The mending required considerable EVA, or spacewalks. In their first spacewalk the astronauts reported that solar array wing 2 and most of the meteoroid shield were gone. Solar array wing 1 appeared intact, but a metal strap held it closed. Later analysis determined that a design flaw, an opening at the top of the meteoroid shield, had allowed air to enter between the station's skin and the shield during ascent. This had created the overpressure that had ripped away the shield, which in turn had snagged and torn away solar array wing 2.

To carry out repairs, the astronauts moved the Skylab 2 command module close to the jammed array. Paul Weitz then stood with his upper body through the hatch and assembled a fifteen-foot pole with a shepherd's hook on the end from three five-foot sections handed to him by Joe Kerwin. He hooked and pulled on the array while Kerwin gripped his legs. Conrad had to hold the Apollo spacecraft steady, because Weitz's efforts pulled it toward the workshop. Weitz replaced the hook with a universal-prying tool when the strap did not budge, but to no avail. Their efforts thwarted, the astronauts docked with Skylab and closed out a twenty-two-hour day. Conrad was blunt about the likelihood of freeing array wing 1: "We ain't gonna do it with the tools we got." The Skylab 2 crew then deployed a solar shield parasol through a small airlock designed for use in deploying scientific experiments, with the intention of shading the spacecraft from solar heat.[48]

The crew commenced their research program on May 29, but the four Apollo Telescope Mount's "windmill" arrays proved insufficient to power the station. But because of the deployment of the parasol sunshade, Skylab cooled sufficiently so that by June 4, 1973, the workshop proved habitable and the crew occupied the it full time. Meanwhile, a team on the ground led by astronaut Rusty Schweickart developed an EVA solar array repair procedure. The astronauts who were in space fabricated tools from onboard materials to carry out his instructions, assembling six five-foot rods with a cable cutter attached to the end. They then tied twenty feet of rope to pull the cutter. This permitted the astronauts to operate the cutter from afar. This tool, along with a Beam Erection Tether, enabled astronauts Conrad and Kerwin to free the jammed solar array and increase power to the workshop during a critical June 7 EVA of three hours and twenty-five minutes. Deploying the array had not been an easy task. As Conrad later described it: "I was facing away from it, heaving with all my might, and Joe was also heaving with all his might when it let go and both of us took off. By the time we got settled down . . . those panels were out as far as they were going to go."[49]

The crew then carried out its mission as planned, performing experiments for twenty-eight days. In orbit the astronauts conducted solar astronomy and Earth resources experiments, medical studies, and five student experiments. The Skylab 2 mission especially concentrated on life sciences and studied the cardiovascular, musculoskeletal, hematological, vestibular, metabolic, and endocrine systems of the body. Its crew also explored the heart's electrical forces during the application of lower body negative pressure, and studied microgravity-induced changes in astronauts during and after flight. In addition, the astronauts used a metabolic analyzer, a device that measures oxygen consumption, to study respiratory responses to bicycle exercise to determine how the capacity to do physical work in space differed from that on the ground. At a fundamental level, the Skylab 2 mission initiated a comprehensive medical research program that extended the knowledge gained in the Mercury, Gemini, and Apollo programs and provided scientific data that would later be used for planning longer duration flights.[50]

This first crew made 404 orbits and carried out experiments for 392 hours, in the process making three EVAs totaling 6 hours and 20 minutes. This doubled the previous length of time in space, making Skylab 2 a major milestone for NASA. That first group of astronauts returned to Earth on June 22, 1973, and two other Skylab missions followed.

Skylab 3 was launched using Apollo hardware on July 28, 1973, and its mission lasted 59 days, 11 hours, and 9 minutes. A total of 1,084.7 astronaut-utilization hours were tallied by Skylab 3 astronauts performing scientific experiments in the areas of life sciences, solar physics, Earth resources, material science, and various student projects. Skylab 4, the last mission on the workshop, was launched on November 16, 1973, and remained in orbit for eighty-four days. This crew had difficulty keeping up with the rigorous schedule made for them by scientists on the ground, and by the end of the year their anger at being overtaxed was apparent to all. NASA officials then relaxed the work schedule to allow somewhat more free time for the astronauts. This paid huge dividends, as productivity increased. By the time the crew returned to Earth in February 1974, it had taken twice as many pictures of the Sun as expected and had conducted thirty-nine Earth surveys. At the conclusion of Skylab 4, the orbital workshop was powered down with the intention that it would be visited again in four years.

To rescue Skylab, NASA constructed a large solar shade. Here, two seamstresses stitch together a sunshade. NASA engineers and scientists worked around the clock on the emergency repair procedure. The Skylab crew and the repair kits were launched just eleven days after the incident, and the crew successfully deployed the twin-pole sail parasol sunshade during their EVA (extravehicular activity) the next day. (NASA photo, no. MSFC-7040525)

When Skylab lost its micrometeoroid shield during the launch, a metal strap became tangled over one of the folded solar array panels. This photograph shows astronauts Schweickart and Gibson using various cutting tools and methods in Marshall Space Flight Center's Neutral Buoyancy Simulator (NBS), a huge water tank that simulated the weightless environment of space. Extensive testing and many hours of practice in simulators such as the NBS tank helped prepare crew members for EVA performance. (NASA photo, no. MSFC-7040570)

During the three missions, a total of nine astronauts had occupied the Skylab workshop for a total of 171 days and thirteen hours. Skylab proved significant for offering an enormous expansion of total hours in space and total hours spent in performance of EVA under microgravity conditions. These combined totals overshadowed all of the world's previous spaceflights up to that time. NASA was delighted with the scientific knowledge gained about long-duration spaceflight during the Skylab program, despite the workshop's early and reoccurring mechanical difficulties. It was the site of nearly three hundred scientific and technical experiments.[51]

Following the final occupied phase of the Skylab mission, ground controllers performed some engineering tests of certain systems—tests that ground personnel were reluctant to do while astronauts were aboard. They used the Apollo spacecraft boosters to lift Skylab to a higher orbit and shut down its systems. They took these steps on the understanding that sometime before the end of Skylab's nine-year service life, a Space Shuttle would dock with the laboratory and reactivate it for at least one more mission.

Unfortunately, this proved impossible. Many at NASA considered Skylab not worth keeping, thinking that they would be able to build a bigger and better space station, and therefore there was no reason to keep it in orbit. As former Utah senator Jake Garn recalled in 2000: "In hindsight it was a bad mistake because actually Skylab was big, had a lot of room in it. . . . If people at that time had envisioned that we still wouldn't have a space station in 2000, the decision might have been different. And it was hard to argue at the time to keep it up there in light of the plans for a successor."[52]

Additionally, it was expected that Skylab would remain in orbit eight to ten years, by which time NASA might be able to reactivate it. But in the fall of 1977, NASA engineers realized that Skylab had entered a rapidly decaying orbit caused by the solar wind and that it would reenter the Earth's atmosphere before 1980. Meantime, the shuttle program had to be delayed because of cost and technical difficulties, pushing back the first flight from its planned date in 1978 until 1981.

An increase in solar activity steadily pushed Skylab into successively lower orbits, until it began to reenter the atmosphere. NASA steered the orbital workshop somewhat to prevent reentry over populated areas, but on July 11, 1979, Skylab finally hit the Earth. The debris stretched from the southeastern Indian Ocean across a sparsely populated section of western Australia. NASA and the U.S. space program took criticism for this development, ranging from the sale of hardhats as "Skylab Survival Kits" to serious questions about the propriety of spaceflight altogether, if people were likely to be killed by falling objects. It was an inauspicious ending to the first American space station, not one that its originators had envisioned. But it further whetted the appetite of NASA leaders for a full-fledged station.

The crew of Skylab 4 took this picture of Skylab in orbit, showing the sunshade still in place. Skylab 4 was the last of its series. (NASA photo, no. MSFC-7449862)

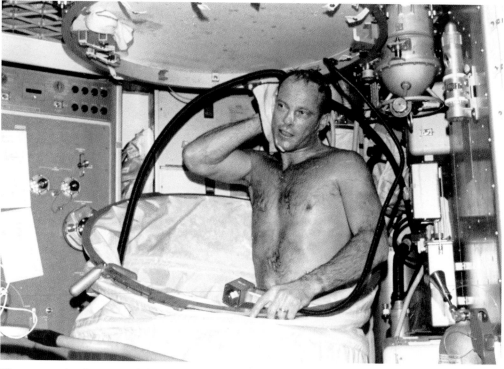

The zero-gravity shower on Skylab taxed the design capabilities. Here Jack R. Lousma, the Skylab 3 pilot, takes a hot bath in the crew quarters of the orbital workshop. (NASA photo, no. SL3-108-1295)

An Assessment of Skylab

More than anyone realized at the time, Skylab proved to be a remarkably successful project during the 1973–74 period. It gave NASA medical specialists and engineers some of the most important data they had ever received to argue for a more permanent home in space, and it proved—for tours of duty not exceeding ninety days—that microgravity conditions had no detrimental effect on human beings. The standard work on space medicine stated the following:

> Skylab provided a wealth of biomedical data concerning the health and physiological responses of humans performing normal work activities and using countermeasures during long-term space missions. Skylab data were particularly useful in differentiating self-

This photograph shows astronaut Owen Garriott atop the Apollo Telescope Mount, removing a film magazine (white box) from one of Skylab's solar telescopes during an EVA in the second manned Skylab mission, Skylab 3. A long boom transported it back into the waiting hands of another crew member at the airlock door below. During the operation, Garriott, film, boom, and Skylab were 270 miles high and speeding around the Earth at 26,000 miles per hour. Because they moved together with no wind resistance, there was little sense of motion. (NASA photo, no. MSFC-7047445)

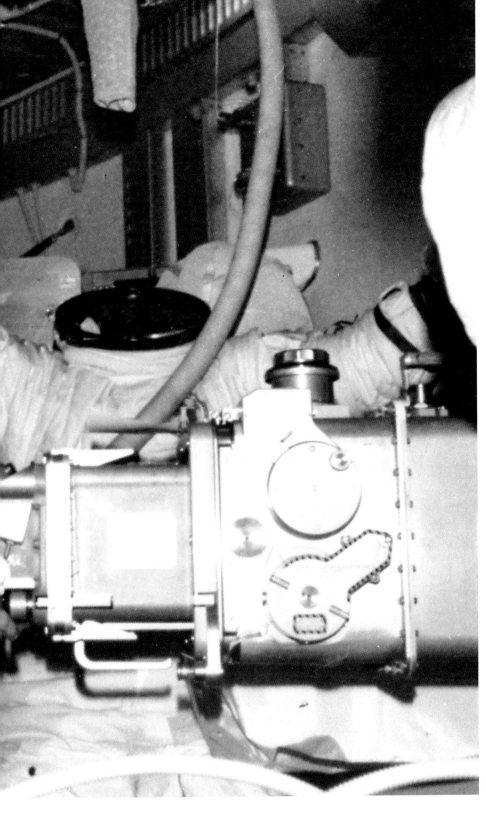

Skylab 3 astronaut Alan Bean operates the Ultraviolet (UV) Stellar Astronomy experiment in the Skylab Airlock Module. The S019, a camera with a prism for UV star photography, studied the spectra of early stars and galaxies. (NASA photo, no. MSFC-0101912)

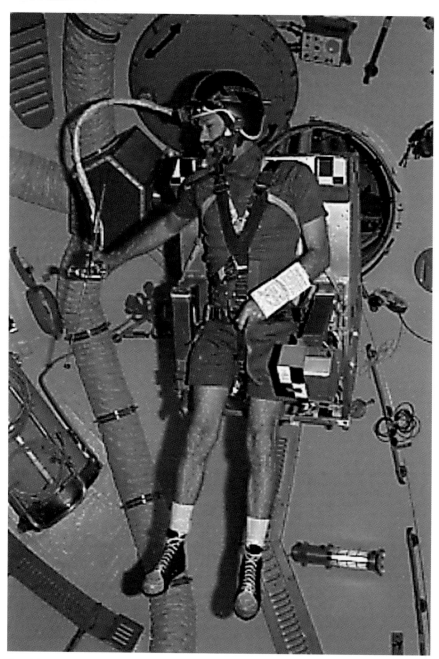

Gerald Carr, an astronaut aboard Skylab 4, tests Astronaut Maneuvering Equipment by flying it around under weightless conditions in the orbital workshop. The M509 experiment was an operational study to evaluate and conduct an in-orbit verification of the utility of various maneuvering techniques to assist astronauts in performing tasks that were representative of future EVA requirements. (NASA photo, no. MSFC-745054I)

limiting physiological changes from those that continue throughout exposure to weight-lessness. This information has guided ground-based research as well as in-flight studies that seek to characterize human responses to the stresses of space.[53]

Skylab also proved the viability of long-term human spaceflight. As scientists concluded, "with sufficient attention to such issues as food service, waste management, and sleep arrangements, a spacecraft could provide satisfactory living and working quarters for long periods."[54] The mission yielded a wealth of scientific data about the Earth as well. This impressed even the Soviets, who discussed it at length with the NASA engineers who were planning with them the Apollo-Soyuz Test Project that would be flown in 1975. As reported in 1974: "Skylab/Apollo experiments have given us the experience to permit the design of a remote sensing satellite appropriate to an economically viable, operational system." This laid the groundwork for stunning Earth science discoveries in the latter part of the twentieth century.[55] As for business possibilities, Skylab research showed that the "space environment has some unique effects on materials processing. It is potentially possible to translate these effects into tangible benefits such as commercial products produced in space."[56]

Perhaps most important, Skylab confirmed the role of humans in space, a debate that had raged since the electronics revolution of the 1960s. More than just an end in itself, the program reinforced the notion that having humans in space was crucial to the future of spaceflight. The Skylab 2 crew's ability to repair much of the damage done during the workshop's launch and make it into a usable facility went far toward demonstrating the value of the astronaut as a "worker" rather than a "hero." A similar discovery resulted for the Soviet Union, as cosmonauts began to earn their keep in a series of Salyut space stations in the early 1970s. In carrying out their space duties, they quashed the authority of arguments that denigrated human spaceflight in favor of robotic missions.

Since the late 1970s, the most reasoned commentary on this subject has recognized the value to space exploration of both human and robotic missions. Scientist Paul D. Spudis summarized this position in a 1999 article:

Judicious use of robots and unmanned spacecraft can reduce the risk and increase the effectiveness of planetary exploration. But robots will never be replacements for people. Some scientists believe that artificial intelligence software may enhance the capabilities of unmanned probes, but so far those capabilities fall far short of what is required for even the most rudimentary forms of field study.

To answer the question "Humans or robots?" one must first define the task. If space exploration is about going to new worlds and understanding the universe in ever-increasing detail, then both robots and humans will be needed. The strengths of each partner make up for the other's weaknesses. To use only one technique is to deprive ourselves of the best of both worlds: the intelligence and flexibility of human participation and the beneficial use of robotic assistance.[57]

The debate has continued, but Skylab showed the utility of humans in orbit, aided in collecting critical scientific data, and established more than any other project in the space program the value of an orbital facility as a future base camp to the stars.

The Soviet Union's Space Station Effort

Even as the United States involved itself in preparing Skylab for launch, the Soviet Union built and flew its own space station.[58] Like the Americans, Soviet space advocates had pressed for the development of a station for that nation's program from almost the beginning of the space age. As early as 1962, Soviet engineers proposed a space station comprised of modules launched separately and brought together in orbit. Even as the United States was still completing Project Apollo, in 1971 the Soviet Union launched its first space station. Its first-generation station had one docking port and could not be resupplied or refueled. There were two types: highly secret Almaz military stations and a public set of Salyut civilian stations.

The Soviet Salyut 1 in an assembly shop in 1971. (APN photo, no. A71-11017)

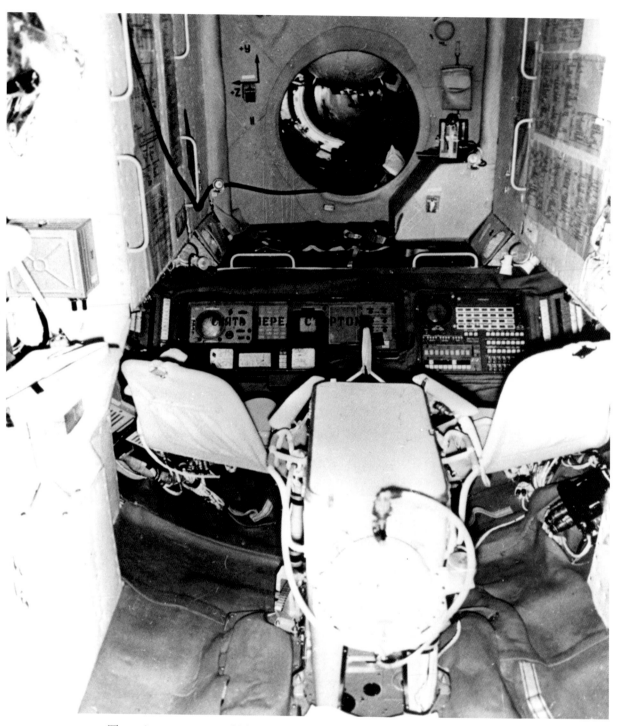

The main compartment of Salyut I in 1971. Seen in the center are the control panel and crew chairs, with the hatch in the background leading to the passageway between Salyut and the Soyuz ferry ship. (APN photo, no. A71-11047)

The Almaz program was the first approved. When proposed in 1964, it had three parts: the military surveillance space station, Transport Logistics Spacecraft for delivering soldier-cosmonauts and cargo, and Proton rockets for launching both. Although these spacecraft were built, none was used as originally planned. Military reconnaissance was Almaz's primary goal, but other investigations were also envisioned. The space station weighed about 9,500 pounds (or 18,000 kg) and would serve a crew of two to three cosmonauts brought to it by a separate space vehicle. The crew would conduct research and then use the transport vehicle's reusable reentry capsule to return to the USSR. By 1970 the station was ready for launch, but the transport vehicle's development was delayed, partly by the inability to "man-rate" the Proton rocket, or make it reliable enough for human use. (*Man-rating* is a term for measuring reliability. It has to be successful during launch 99.9999 times out of 100.) During this delay, it was decided to modify the station and to use the Soyuz spacecraft as a crew transport vehicle. The Soviets called the modified program Salyut.[59]

These changes represented a direct effort to counter the American success with Apollo. They emphasized a civilian program so that the Soviet Union could recover a measure of international prestige with a spectacular public success. Salyut 1, the first space station in history, reached orbit atop a Proton rocket on April 19, 1971.

Those early first-generation stations were plagued by failures. The crew of Soyuz 10, the first spacecraft sent to Salyut 1, was unable to enter the station because of a docking mechanism problem. The Soyuz 11 crew of three lived aboard Salyut 1 for three weeks, but died during return to Earth because air escaped from their spacecraft. After that, three first-generation stations failed to reach orbit or broke up in orbit before crews could reach them. The second failed station was Salyut 2, the first Almaz military station to fly. The Soviets recovered rapidly from these failures. Salyut 3, Salyut 4, and Salyut 5 supported a total of five crews. In addition to military surveillance and scientific and industrial experiments, the cosmonauts performed engineering tests to help develop the second-generation space stations.[60]

With the second-generation stations, the Soviet program evolved from short-duration to long-duration stays. These stations had two docking ports to permit refueling and resupply by automated Progress freighters derived from Soyuz spacecraft. Progress docked automatically at the aft port and was then opened and unloaded by cosmonauts on the station. Transfer of fuel took place automatically, under supervision from the ground. A second docking port also meant long-duration resident crews could receive visitors. Visiting crews often included cosmonaut-researchers from Soviet bloc countries or countries sympathetic to the Soviet Union. For example, Vladimir Remek of Czechoslovakia, the first space traveler not from the United States or the Soviet Union, visited Salyut 6 in 1978.[61]

The cosmonauts developed strict procedures for rotating equipment at these stations.

101

SKYLAB

AND THE

SALYUTS:

PRELIMINARY

SPACE STATIONS

This artist's conception depicts the docking of Soyuz 11 to the Salyut 1 Soviet space station. (APN photo, no. A71-10827)

A Soyuz capsule on an R-7 launcher in 1970s. (NASA photo)

Rotating crews often traded their Soyuz spacecraft for the one already docked at the station, because Soyuz had only a limited lifetime in orbit. They incrementally extended their service times in orbit from 60 to 90 days for the Soyuz Ferry to more than 180 days for the Soyuz-TM. The first of these extended-duration stays took place on Salyut 6, a civilian station that orbited between 1977 and 1982. It received a total of sixteen cosmonaut crews, including six long-duration crews, and hosted cosmonauts from Hungary, Poland, Romania, Cuba, Mongolia, Vietnam, and East Germany. An experimental transport logistics spacecraft called Cosmos 1267 docked with Salyut 6 in 1982. It proved that large modules could dock automatically with space stations, a major step in technology, and the procedures pioneered there are in use to the present.

The Soviet Union launched Salyut 7 in 1982, and although it remained in orbit until 1991, its last crew flew in 1986. A near-twin of Salyut 6, this station was home to ten cosmonaut crews, including six long-duration crews, one of which set a record for orbiting 237 days. The Soviet Union expanded its crew complement significantly during the operational life of Salyut 7, inviting France and India to send astronauts and flying the first female cosmonaut since 1963. Salyut 7 was abandoned in 1986, when the Soviet Union began operating its first long-duration space station, Mir. It was not until 1991 that Salyut 7 reentered Earth's atmosphere, over Argentina.

The central question of all historical studies is, of course, why any of this matters. In this case the dream of a permanent presence in space, made sustainable by a vehicle providing routine access at an affordable price, has driven space exploration advocates since at least the beginning of the twentieth century. All of the spacefaring nations of the world have accepted that paradigm as the raison d'être of their programs in the latter twentieth century. The dream drove the United States to develop and fly Skylab in the 1960s and 1970s, and the Soviet Union to build its Salyut series of stations in the same time frame. Ultimately, everyone working in the program agreed, these would eventually lead to a permanent human presence in space.

At the same time, this vision has not always been consistent with many of the elements of political reality in the United States. Numerous questions abounded in the aftermath of Apollo concerning the need for aggressive human exploration of the solar system and the desirability of colonization on other worlds, an endeavor in which a space station would logically serve as base camp. But, as political scientist Dwayne A. Day wrote, such a project "implies that a long-range human space plan is necessary for the nation without justifying that belief. Political decision-makers have rarely agreed with the view that a long range plan for the human exploration of space is as necessary as—say—a long range plan for attacking poverty or developing a strategic deterrent. Space is not viewed by many politicians as a 'problem' but is at best an opportunity and at worst a lux-

105

SKYLAB

AND THE

SALYUTS:

PRELIMINARY

SPACE STATIONS

The Salyut 7 in Earth orbit, mid-1980s, with a Soyuz T ferry craft attached. (NASA photo)

1971–74	1975–76	1977	1978	1979	1980
Soyuz 10 Apr. 22, 1971 *1:23:46*	**Soyuz 17** Jan. 10, 1975 *29:13:20*	**Soyuz 24** Feb. 7, 1977 *17:17:23*	**Soyuz 27** Jan. 10, 1978 *64:22:53*	**Soyuz 32** Feb. 25, 1979 *108:4:24*	**Soyuz 35** Apr. 9, 1980 *55:1:29*
Soyuz 11 June 6, 1971 *23:18:22*	**Soyuz 18** May 24, 1975 *62:23:20*	**Soyuz 25** Oct. 9, 1977 *2:0:46*	**Soyuz 28** Mar. 2, 1978 *7:22:17*	**Soyuz 33** Apr. 10, 1979 *1:23:1*	**Soyuz 36** May 26, 1980 *65:20:54*
Soyuz 14 July 3, 1974 *15:17:30*	**Soyuz 21** July 6, 1976 *48:1:32*	**Soyuz 26** Dec. 10, 1977 *37:10:6*	**Soyuz 29** June 15, 1978 *9:15:23*	**Soyuz 34** June 6, 1979 *7:18:17*	**Soyuz T-2** June 5, 1980 *3:22:21*
Soyuz 15 Aug. 26, 1974 *2:0:12*	**Soyuz 23** Oct. 14, 1976 *2:0:6*		**Soyuz 30** June 27, 1978 *7:22:4*		
			Soyuz 31 Aug. 26, 1978 *67:20:14*		

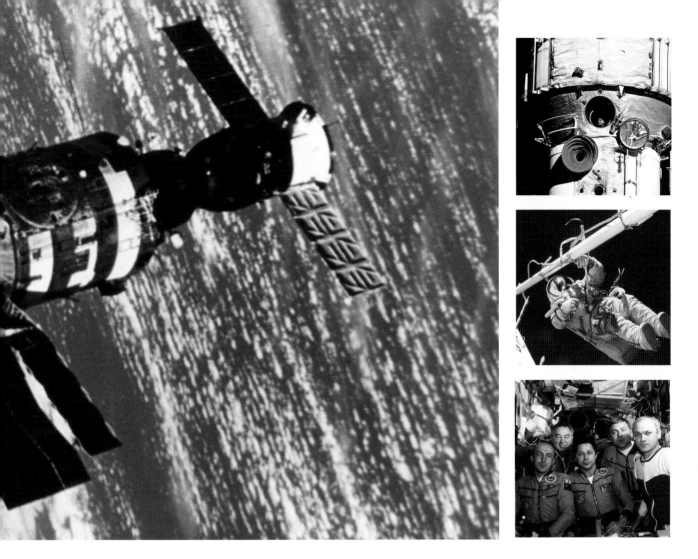

1980	1981	1982	1983	1984	1985-86
Soyuz 37 **July 23, 1980** *79:15:17*	**Soyuz T-4** **Mar. 12, 1981** *74:18:38*	**Soyuz T-5** **May 13, 1982** *211:9:5*	**Soyuz T-8** **Apr. 20, 1983** *2:0:18*	**Soyuz T-10** **Feb. 8, 1984** *62:22:43*	**Soyuz T-13** **June 5, 1985** *112:3:12*
Soyuz 38 **Sep. 18, 1980** *7:20:43*	**Soyuz 39** **Mar. 22, 1981** *7:20:43*	**Soyuz T-6** **June 24, 1982** *7:21:51*	**Soyuz T-9** **June 28, 1983** *149:9:46*	**Soyuz T-11** **Apr. 3, 1984** *181:21:48*	**Soyuz T-14** **Sep. 17, 1985** *64:21:52*
Soyuz T-3 **Nov. 27, 1980** *12:19:8*	**Soyuz 40** **May 14, 1981** *7:20:41*	**Soyuz T-7** **Aug. 19, 1982** *7:21:52*		**Soyuz T-12** **July 17, 1984** *11:19:14*	**Soyuz T-15** **Mar. 13, 1986** *125:1:1*

SALYUT 1-7 HUMAN
SPACEFLIGHTS, 1971–86
(DAYS, HOURS, MINUTES IN ORBIT)

Source: Aeronautics and Space Report of the President, 1999 Activities (Washington, D.C.: NASA Annual Report, October 2001), appendix C.

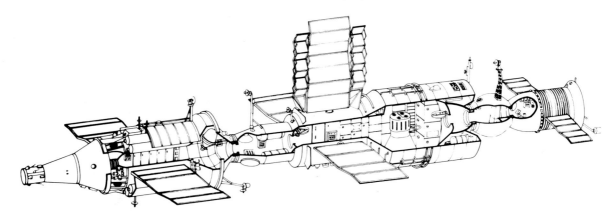

A conceptual drawing of the interior of the Kosmos 1443-type
Module, Salyut 7, and a Soyuz T ferry craft. (NASA photo)

ury."[62] Most important, the high cost of conducting space exploration continues to come quickly
into any discussion of the idea.

Space has always had the ability to excite and inspire humanity, just as exploration of the
world beyond Europe in the fifteenth and sixteenth centuries inspired and excited those nations.
Like those earlier explorations, it holds the allure of discovery of a vast unknown awaiting human
assimilation, an allure particularly appealing to a society such as the United States, so heavily influenced by territorial expansion. In many respects, space exploration represents what the 1960s
popular culture television phenomenon of *Star Trek* dubbed it, the "final frontier." And it is the
concept of a space station as a permanent presence in Earth orbit that has driven the major program of NASA from 1984 to the present.

Soviet cosmonauts aboard the Salyut 7 space station in Earth orbit. *From left:* V. Vasyutin, G. Grachkp, V. Savinykh, A. Volker, and V. Dzhanibatov. (NASA photo, no. 88-HC-485)

CHAPTER 4

The Strange Career of Space Station Freedom

The history of Space Station Freedom may have more in common with that of the fictitious vampire Dracula than with a space project: many thought it neither dead nor alive and constantly tried to drive a stake through its heart. With Skylab gone from the scene after 1979, and the beginning flights of the Space Shuttle in 1981, NASA returned during the early 1980s to its quest for a real space station as a site of orbital research and a jumping-off point to the planets. In 1984 NASA leaders persuaded President Ronald Reagan, against the wishes of many advisers, to endorse the building of a permanently occupied space station. The next year the agency came forward with designs for an $8 billion dual-keel space station configuration,

As the Space Shuttle was being built, in the mid-1970s NASA began pursuing a new space station program. Many of the designs sought to expand on the forthcoming Shuttle, seeking dual use for its hardware. For example, a 1977 concept proposed using the spent main engine tank of the Space Shuttle as a crew habitat and baseline for a permanent presence in space. The proposal called for 2,000 cubic feet of the forward area of the main tank to be prepared on the ground for crew habitation in orbit. While living in this area, a team of three would then construct additional living and working space in the tank's remaining 17,500 cubic feet. The Space Shuttle entered service in 1981. (NASA photo, no. 77-H-144)

A more adventurous 1977 approach that also used Space Shuttle hardware proposed a "spider" concept for unwinding a solar array from the spent main fuel tank. This device would be capable of forming and assembling a structure in one integrated operation. With such an unwinding of a solar array, the main engine tank could then become a control center for space operations, a crew habitat for Shuttle astronauts, and a focal point for space operations, including missions to the Moon and Mars. (NASA photo, no. 77-H-204)

to which were attached a large solar power plant and several modules for microgravity experimentation, life science, technical activities, and habitation. This station also had the capacity for significant expansion through the addition of other modules.

From the outset, both the Reagan administration and NASA intended Space Station Freedom, as it was called, to be an international program, and over the course of the 1980s several nations entered the program. Even so, almost from the outset Space Station Freedom was controversial. Most of the debate centered on its costs versus its benefits. Redesigns of Space Station Freedom took place in 1990, 1991, 1992, and 1993. Each time the project got smaller, less capable of accomplishing the broad projects originally envisioned for it, less costly, and more controversial. As costs were reduced, capabilities also had to diminish, and political leaders who had once supported the program increasingly questioned its viability. It was a seemingly endless circle, and political wits wondered when the dog would wise up and stop chasing its tail. Some leaders suggested that the nation, NASA, and the overall space exploration effort would be better off if Freedom were terminated.

In the latter 1980s and early 1990s, a parade of space station managers and NASA administrators, each of them honest in their attempts to rescue the program, wrestled with Freedom and lost. They faced on one side politicians who demanded that the government contracts for the station—itself a major cause of the overall cost growth—be maintained; on the other side station users insisted that Freedom's capabilities be maintained. On both sides people demanded that costs be reduced. The incompatibility of these various exigencies ensured that station program management was a task not without difficulties. Finally, in 1993, the program was transformed into the present International Space Station. This chapter discusses that process.

THE SPACE STATION DECISION

Skylab was a genuine success for NASA, and most people recognized it as such. Only its coordinated fiery demise in 1979 raised negative publicity for the agency and questions about the need for a space station. But the dream did not die in the intentional crash that consumed Skylab. Instead, it metamorphosed into an even more expansive effort, one that would provide the long-desired permanent human presence in Earth orbit, the single most significant step to be taken for humanity's ultimate task of leaving this planet.

Even as the space shuttle development program proceeded during the middle part of the 1970s, NASA's planning staff prepared space station concept studies with the objective of being prepared to bring them forward should a political opportunity present itself.[1] In early 1975, for example, the NASA administrator established an Outlook for Space Study Group and chartered it to conduct a review of potential space activities for the last quarter of the century. The resulting study, *Outlook for Space,* stated that a station was "the next logical requirement for proceeding along the continuum of space objectives."[2] It came as no surprise to those within the community, therefore, once NASA began flying the Space Shuttle in 1981, that administrator James M. Beggs pressed for approval to develop a space station as the "next logical step" in human exploration of the region beyond Earth's atmosphere.[3]

Beggs proved a talented and tireless proponent of a space station program in the period between 1981 and 1984. At his Senate confirmation hearing in June 1981, members of Congress asked him what should be the next major U.S. undertaking in space. He replied without hesitation that "it seems to me that the next step is a space station." Beggs believed that this would

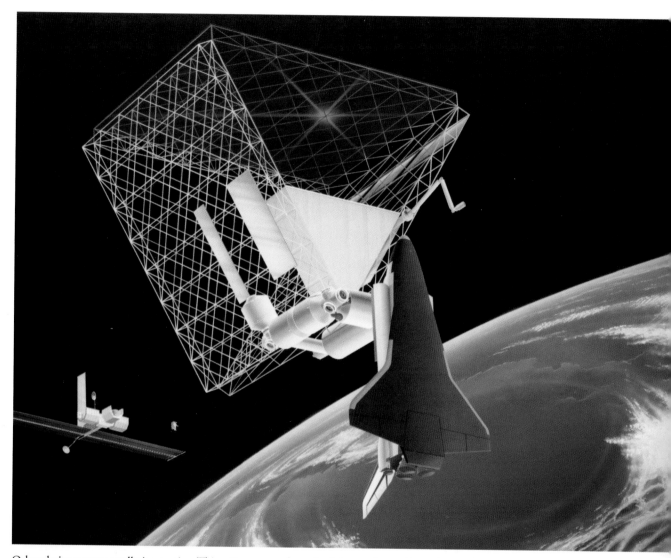

Other designs were equally innovative. This 1984 concept from the Johnson Space Center featured a rigid framework "roof," covered with fixed solar array cells. Always pointed toward the Sun, its panels would generate about 120 kilowatts of electricity. Rigid V-shaped beams would hold five modules for living and laboratory space, and additional external areas would be available for instruments and other facilities. (NASA photo, no. 84-HC-17)

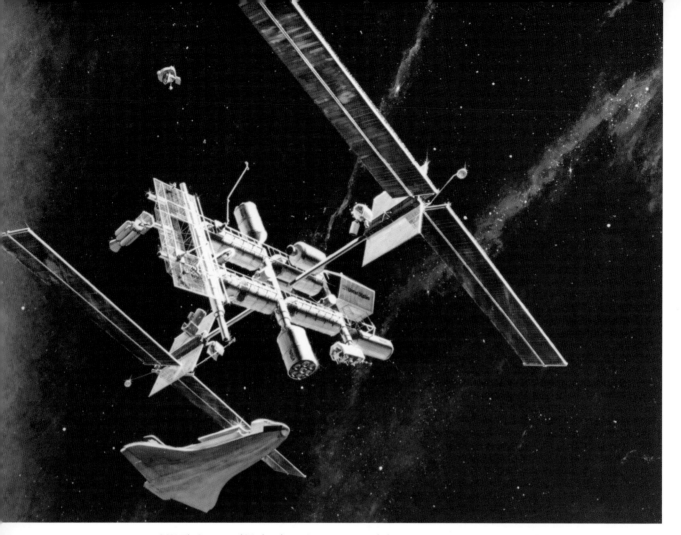

TRW's Space and Technology Group proposed this 1984 concept. It consisted of dual solar array sets, logistics modules, command and habitat modules, attached pallets for scientific experiments, and orbital transfer vehicles. (NASA photo, no. S-84-29874)

serve all of the purposes developed throughout the twentieth century, including that of a stepping-stone to other bodies in the solar system.[4] Under his direction NASA waged an intense, and ultimately successful, campaign to gain presidential approval to develop a large, permanently occupied space station.

From the very beginning of his quest, Beggs received the support of a few highly placed Reagan administration officials. When Ronald Reagan entered the White House following the 1980 election, his space policy transition team, led by former NASA deputy administrator George M. Low, recommended an aggressive exploration agenda that included as a centerpiece the long-desired space station. Low's report commented: "The year 1980 finds NASA in an untenable position. . . . This unhealthy state of affairs can only be rectified by a conscious decision. Continuation of the prior administration's low level of interest and lack of clear direction would result in an unconscionable waste of human and financial resources." The report stopped just short of advocating a space station as critical for all future spaceflight endeavors.[5]

Between 1981 and the eventual decision in late 1983, Beggs carefully laid the foundation for Reagan administration approval. His approach accentuated making the Space Shuttle operational in the first two years of the administration, and then urging the station as the next step in the "normalization" of spaceflight in low Earth orbit. As two of Beggs's chief lieutenants wrote in August 1981, the space station "should become the major new goal of NASA and, some time during the next two years, the President should be persuaded to issue a statement proclaiming a national commitment to that effect." The arguments justifying this commitment, Beggs believed, "range from national security (i.e., arms control verification, military surveillance) to the improvement of space operations (i.e., satellite maintenance on orbit and other things of this kind)." Without it, NASA leaders believed, "exploration will be de-emphasized somewhat until we have a Space Station that can serve as a base for the launching of a new generation of planetary exploration spacecraft."[6]

To prepare the way Beggs ordered a set of intensive internal and contractor studies, with a particular focus on identifying in detail the missions that a space station might perform. He also pursued a dual strategy for winning approval. One path emphasized cooperation with other government agencies and external constituencies to build a broad coalition in support of the station, while the other went straight to the top to convince President Ronald Reagan that it was critical to American interests to proceed with the program.[7] In the first instance, Beggs worked within the administration's interagency policy-coordinating entity to win consensus. Under Reagan the key organization was the Senior Interagency Group (SIG) for Space, chaired by the president's assistant for national security, which oversaw space policy formulation. Beggs persuaded the members of SIG (Space) to support a study that would provide the basis for a presidential decision on whether to proceed with a space station. Among the questions that SIG (Space) had were these:

- How will a manned Space Station contribute to the maintenance of U.S. space leadership and to the other goals contained in our National Space Policy?
- How will a manned Space Station best fulfill national and international requirements versus other means of satisfying them?
- What are the national security implications of a manned Space Station?
- What are the foreign policy implications of a manned Space Station?
- What is the overall economic and social impact of a manned Space Station?[8]

It became clear early in the study cycle that regardless of the findings, the majority of members in SIG (Space) would not support a decision to build a space station. When Beggs tried to justify it as the next logical step in space exploration, SIG (Space) members asked, "step toward what?" His pointing out the many missions proposed for a permanently occupied orbiting laboratory, however, failed to convince them. Knowing it would only excite greater opposition, during the SIG (Space) debates over the space station in the 1982–83 time frame, he refused to identify the station with the ambitious goal of preparing for human journeys to Mars, despite pressure from presidential science adviser George A. Keyworth II to do so. Beggs said that the negative response to the 1969 Space Task Group recommendation for Mars exploration had taught him a strong lesson about rushing too far ahead of public approval. He judged that the time was not propitious for linking station approval to such a visionary objective.

Even so, he could not persuade the members of SIG (Space) that the space station held any value as a major government project. In particular the Department of Defense, led by Secretary of Defense Caspar Weinberger, argued against the project. As an example of this position, on June 20, 1983, the Department of Defense informed Robert McFarlane, Reagan's deputy assistant for National Security Affairs, that the military could not identify a mission that could be uniquely fulfilled by a space station of any type:

> Further . . . no projected requirements could be identified that would be significantly enhanced by the development of a space station. . . . Studies to date have identified no unique, cost effective contributions that man-in-the-loop [human spaceflight rather than robotic spacecraft] can make to the execution of military missions such as surveillance, navigation and communications. Further, considering the cost of developing and procuring one or more space stations and the difficulty in making a space station survivable, questions are raised concerning the reliance that could be placed on the availability of a space station in conflict.[9]

As SIG (Space)'s position became clear to him, Beggs began pursuing even more diligently his second course of directly trying to convince the president. He believed that President Reagan would approve the space station anyway, should he be presented with the opportunity. All he had to do, Beggs believed, was to get the project before him. Effectively using his friends in the White House, he succeeded in having the issue of building a space station added to the agenda of a December 1, 1983, Cabinet Council on Commerce and Trade where national security participants were few.[10]

The NASA presentation at this meeting, with the president in attendance, asked for a decision to proceed with the space station program. Beggs stressed its potential contribution to

On December 1, 1983, NASA administrator James M. Beggs proposed to a White House Cabinet Council of Commerce and Trade meeting the building of a space station. Attendees included Office of Management and Budget director David Stockman *(second from left)*, Vice President George Bush *(partially hidden, fourth from left)*, the president's science adviser George Keyworth *(at end of table next to overhead projector)*, and President Ronald Reagan *(second from right)*. Beggs emerged from the meeting with a tentative commitment to proceed with a station. (NASA photo)

the leadership of the United States on the world stage. He knew that Ronald Reagan had long been concerned with a perceived withering of American prestige vis-à-vis the Soviet Union. The station, he argued, would help to quell that declension, and it would represent a continuation of American strength in spaceflight that would "dominate the space environment for twenty years." The presentation discussed potential commercial and scientific windfalls from station-based activities, but Beggs remained silent on the role traditionally envisioned for a station, the jumping-off point from which humanity would journey on. As the punch line for the briefing, Beggs showed a photo of a Salyut space station overflying the United States. He stressed the fact that the Soviet Union already had this modest station and was planning a larger orbital facility. Should not the United States have one as well?

Beggs asked this question knowing full well that Reagan viewed the Soviet Union as an "evil empire" bent on the destruction of the United States. Driving home his point, he continued, "What might the Soviets be planning to do to the United States from this new high ground?" In concluding his masterful presentation, James Beggs told the president and others in the Cabinet Room, "The time to start a space station is now."[11] Reagan agreed.

It would be easy to assess this decision-making effort as fundamentally flawed. One might conclude that Beggs's circumvention of SIG (Space) displayed rather Byzantine power politics. One might just as easily note an overarching deficiency of understanding on the part of the president about the greater needs of the nation in what was admittedly a brief consideration of the issue. But there is also reason to believe that Reagan more thoroughly understood the nuances of power and prestige vis-à-vis the Soviet Union than was largely believed at the time, and that the decision to proceed with a space station fit well into his larger strategy.

There is no question that Reagan viewed the station as a part of a design to defeat the Soviet Union—the evil empire, in his parlance. Plans for a space station raised the possibility of considerable civil-military dialogue concerning the endeavor. Reagan's overarching strategy involved a buildup of U.S. military capabilities; a confrontational approach to foreign policy; assistance to U.S. allies around the world, such as aid to rebels (including Osama bin Laden and the mujahedeen) in Afghanistan; and the development of a multifaceted set of new technological systems. Those ranged from the building of a 600-ship navy and "stealth" aircraft that evaded enemy radar to the Strategic Defense Initiative (SDI), involving the deployment of sophisticated space-based systems that could presumably defeat Soviet missiles launched against the United States. In this context the Reagan administration's objectives for Space Station Freedom, as it was named, necessitated that it push American technological know-how so that spinoff systems might also be used in SDI, and so that it would serve as a rallying point for the nation's allies. As a result, from the outset the space station had to be a high priority international program, for it would serve to enhance the technical competence of the United States and its friends.[12]

All in all, Beggs had been stunningly insightful in sizing up the situation, taking the measure of the president, and circumventing the policy formulation process in Washington. He had, furthermore, proven remarkably resourceful in marshaling supporters in some sectors of the administration, essentially to overcome or circumvent opponents of the space station. President Reagan approved the station program in an Oval Office ceremony a short time later. On January 25, 1984, in his annual State of the Union message, Reagan told Congress and the nation, "sparking economy spurs initiatives, sunrise industries, and makes older ones more competitive." He added:

Nowhere is this more important than our next frontier: space. Nowhere do we so effectively demonstrate our technological leadership and ability to make life better on Earth.

The Space Age is barely a quarter of a century old. But already we've pushed civilization forward with our advances in science and technology. Opportunities and jobs will multiply as we cross new thresholds of knowledge and reach deeper into the unknown. . . .

America has always been greatest when we dared to be great. We can reach for greatness again. We can follow our dreams to distant stars, living and working in space for peaceful, economic, and scientific gain. Tonight, I am directing NASA to develop a permanently manned space station and to do it within a decade. A space station will permit quantum leaps in our research in science, communications, in metals, and in life-saving medicines which could be manufactured only in space.[13]

As they say in sports, "the crowd went wild."

Conducting a Program, or Not

The very public announcement by President Reagan of the commitment to build Space Station Freedom represented the high-water mark of the overall program's support. Clearly the challenges proved enormous, and the trials—political and otherwise—fatiguing, but nothing seemed insurmountable in the first few weeks after the president's speech. After all, NASA had the very public blessing of the president of the United States. But the cheering at the State of the Union announcement masked two overwhelming problems: (1) support for the program proved thin everywhere, since few viewed it as necessary to the welfare of the United States, and funding for it would, therefore, always be difficult to obtain; (2) several members of the Reagan administration, especially at the Pentagon, resented NASA for the manner in which it had gained approval. Not having followed the recognized policy formulation process, those circumvented in space station decision-making believed they had no responsibility to support the effort. Leaders at the Department of Defense especially took every opportunity to question its value.

For example, Secretary of Defense Weinberger informed NASA administrator Beggs that he challenged the whole idea, but he had specific reasons for doing so. "My reservations about your proposal," wrote Weinberger, "relate to cost and impact on the Space Transportation System. . . . In today's constrained fiscal environment, unprogrammed cost growths can only be funded at the expense of other programs. . . . You would not wish to cancel any of your approved civil programs to meet increased funding requirements for a space station any more than we in Defense would like to see our national security budget jeopardized." Weinberger added, "I believe that a major new start of this magnitude would inevitably divert NASA managerial talent and resources from the priority task of making the Space Transportation System fully operational and cost effective. With all our national security space programs commit-

President Ronald Reagan delivering his January 1984 State of the Union address, with Vice President George Bush *(left)* and House of Representatives Speaker Tip O'Neill *(right)* listening. At this address, President Reagan announced the decision to proceed with the building of what was named Space Station Freedom. This speech represented a high point in political support for a space station. (NASA photo)

ted to the Shuttle and dependent on it for their sole access to space, I am sure that you can appreciate my concern in this area."[14] Weinberger would have preferred that the project had never gained approval.

Nonetheless, NASA set to work developing the baseline concept for the space station and organizing to build it. Even before presidential approval, Beggs had an internal NASA Space Station Task Force, under the direction of former NASA flight director John Hodge at Mission Control in Houston, to develop formal plans for the station.[15] From late 1982 to December 1983, this group concentrated on defining missions and user requirements, synthesizing this data, and creating a concept that would satisfy these needs.

The task force found that a station that could accomplish the mission requirements would be expensive, perhaps as much as $12 billion. Beggs asked Hodge and his team to keep the budget to $7 billion, excluding the costs of outfitting, transportation to orbit, and resupply over a five-year period.[16] Sometime in the summer of 1983, Beggs began to use a space station development budget figure of $8 billion. Responding to critics who charged that the real price tag would be closer to $20 billion, Beggs pointed out NASA could purchase a station "by the yard." Since the station was modular, the nation could stop buying and assembling modules at any point above a minimum capability on orbit.[17] Over time the figure of $8 billion gained status as a "line in the sand" for the station's cost, a level chosen to prevent further escalation. One NASA official remembered, "I reached the scream level at about $9 billion," referring to how much politicians appeared willing to spend.[18] As a result, NASA tried to design the project to fit an $8 billion research and development funding profile. For many reasons—some related to tough Washington politics but others to poor management and enormous technical challenges—within five years the projected costs had more than tripled, and the station had become too expensive to fund fully. The national debt had also exploded in the 1980s, and a multitude of other priorities demanded federal funding.[19]

In this environment, convincing Congress that building the space station was a good idea presented some serious obstacles, and its members never willingly gave the program the money necessary to complete the most ambitious program since Apollo. The longstanding NASA conviction that humans would go into space because, as Wernher von Braun once explained, "it is a part of man's destiny" did not persuade many politicians, regardless of party, region served, or background and perspective.[20] Daniel Herman of the Space Station Freedom program office summarized the agency position: "We would still want a permanent presence of man in space, even if it could be proven that functionally everything conceived of today could be done by robots. The reason is that we think it is NASA's charter to essentially prepare for the exploration of space by man in the 21st century." This perspective may not have convinced many members of Congress anyway, but when space scientists offered an alternative position the debate grew more heated. For example, revered astronomer James A. Van Allen, discoverer of the Van Allen radiation belts as a result of the Explorer 1 mission in 1958, complained that "NASA's devotion to the quasi-religious dogma of man's permanent presence in space is so deep-seated as to be immune to the rational discussion of alternatives."[21] Any option that scrapped the space station, however, sent cold shivers down the collective backs of NASA officials.

Even as the political debate raged, in 1984 the agency brought forward a design for a space station configuration that included a solar power plant and modules for microgravity experimentation, life science, technical activities, and habitation. True to the belief that NASA could buy the station "by the yard," it had the capacity for expansion through the addition of other modules.[22] NASA considered five primary concepts for this station—Concept Development Group (CDG) Planar,[23] the Big T, the Delta Truss, the Power Tower, and the Spinner—and narrowed the possibilities down to three: Planar, Delta Truss, and Power Tower. Each had advantages and tradeoffs. The Delta Truss offered a rigid structure, with room for attaching additional payloads and servicing spacecraft. The Planar station featured balanced double-ended solar power arrays, a smaller surface area, and more efficient guidance and control. The Power Tower presented unobstructed and simultaneous Earth and stellar views, afforded easy access, and possessed some future growth potential.[24]

The Power Tower became the concept of choice early on, because it offered the most latitude to design a workable space station given political, funding, and technical limitations. It allowed the satisfying of the broadest range of scientific requirements, and it offered the potential to minimize development costs. Accordingly, the Power Tower configuration best accommodated the proposed mission and made possible future growth in the system if desired. NASA established it as the reference configuration on June 14, 1984.

The Power Tower called for a 300-foot-long latticework "keel" oriented with its axis pointing toward Earth. The pressurized modules would be clustered at the lower end of the keel, so that the "gravity gradient" would stabilize the "heavier" or more massive end. This would reduce the amount of thruster propellant needed to maintain the station. The keel's orientation would therefore remain fixed with respect to the Earth, making it possible to install Earth observation instruments at the base and astronomical ones at the top. A perpendicular truss mounted halfway up would carry the solar panels and thermal radiators, and would give the Space Shuttle unobstructed access to the module cluster. NASA's design of the Power Tower recognized as a given that the Space Shuttle would service the orbital vehicle.[25]

To facilitate the design work, NASA officials divided the effort among four agency centers— Marshall Space Flight Center; Johnson Space Center; Goddard Space Flight Center in Greenbelt, Maryland; and Lewis Research Center in Cleveland, Ohio—in what they referred to as work packages, each with its own contractors designing discrete elements of the station. On September 14, 1984, NASA issued a consolidated Request for Proposals (RFP) for three-year Phase B design studies for Power Tower systems for each of the work packages. Receiving thirteen bids, NASA awarded design contracts on April 1, 1985, with two contractors working in parallel under the supervision of each of the four NASA field centers:

Work Package 1, Marshall Space Flight Center

Contractors: Boeing and Martin Marietta
Total design contract cost: $26 million
Elements:
- a common module for laboratory, habitat, and logistics, deliverable by Space Shuttle
- environmental control and life support systems

Space Station Freedom concepts early hit on two major designs. The first of these was the Dual Keel, shown here. Its solar dynamics power system was slated to generate 50 kilowatts more than other designs could supply. (NASA photo, no. 86-HC-242)

- propulsive systems to maintain station's attitude
- requirements for satellite servicing facilities and orbital transfer

Work Package 2, Johnson Space Center

Contractors: McDonnell Douglas and Rockwell International
Total design contract cost: $30 million
Elements:

- truss structure

The second major design for Space Station Freedom was the Power Tower concept, which NASA selected as the reference configuration in 1984. The Power Tower mounted solar arrays at the top of the trusses and modules, with most experiments taking place at the bottom. (NASA photo, no. 85-HC-244)

- interface between shuttle and station
- guidance, navigation, and control systems
- crew habitability
- thermal control system
- communications and data management systems
- airlock
- EVA systems
- remote manipulator

Work Package 3, Goddard Space Flight Center

Contractors: General Electric and Radio Corporation of America (RCA)
Total design contract cost: $10 million
Elements:
- attached and free-flying payloads
- laboratory systems

Work Package 4, Lewis Research Center

Contractors: Rocketdyne Systems and Thompson, Ramo and Wooldridge (TRW)
Total design contract cost: $6 million
Elements:
- power generation
- energy storage
- distribution systems

A Space Station Program Office would manage the overall effort from the Johnson Space Center. This represented a departure from previous large-scale NASA projects, in which management rested with an office at NASA headquarters.[26]

But the Program Office was a subsidiary of the NASA Office of Space Station in Washington, D.C. The multitude of bureaucracy that these various organizations demonstrated helps to show why the Freedom program failed. The NASA Office of Space Station expected that the results of these three-year Phase B studies—officially designated Definition and Preliminary Design contracts—would yield specifications of sufficient refinement that NASA could then move forward with fabrication, test and verification, system integration, launch, and assembly of the space sta-

As Space Station Freedom evolved from design to design, its orientation changed from a vertical to horizontal line on Earth's horizon. This highly romanticized artist's conception by Alan Chinchar (1991) shows the station in its completed configuration. With Earth as a backdrop, the painting also looks toward the Moon and Mars. (NASA photo, no. 91-HC-717)

tion. In essence the Phase B studies would result in "a total Space Station Program preliminary design with all the necessary cost and schedule estimates and management plans." The belief at the time was that NASA could have the structure in orbit and operational in the early 1990s, within the ten-year time frame that Reagan established in 1984.[27]

KEY FACTORS IN THE EVOLUTION OF
SPACE STATION FREEDOM

The task proved far more difficult than envisioned. Five major factors came to the fore to thwart speedy completion of the design and construction process. Each of these interlocked and played off of each other. The first involved the challenge of developing the technology itself. Designing a major space station and building it proved far more complex and filled with pitfalls than had ever been imagined. Managing this process proved overwhelming, and the complex bureaucracy of the program failed to control it. Second, the political support for the station, never overwhelming, was enormously difficult to maintain as the years passed, as NASA's credibility declined, and as external priorities arose that demanded monetary support. Advocates of the program drained their energies defending it against an onslaught of opposition, repeatedly calling on outside assistance and eventually exhausting what goodwill had once existed. Third, as technical problems mounted and political difficulties compounded, the cost of the project grew. With each rise in the budget projection, a new round of technical and political debate resulted. Fourth, international participation added complexity on an order of magnitude never attempted before. Not even Apollo, despite its enormous intricacies, had had to contend with multiple nationalities with their own political priorities, languages, and styles. Fifth, and perhaps most important, the proposed uses of the space station once on orbit never convinced very many people to support it. Even scientists, who, NASA leaders believed, should have first recognized its potential, wondered if robotic spacecraft could more efficiently perform these missions. And the great cost involved in using the station prohibited its employment for most studies such as microgravity research, materials processing, crystal growth, and the like.

In the first instance, a series of fundamental technical questions required answers, as outlined by historian Adam Gruen:

> How would the modules connect? How long should they be? Should equipment racks be placed on the floors/walls/ceilings or in a central spine running the length of a module? Should a laboratory module be separate from a habitation module, or combined? Where should the airlocks go? How big should the hatches between modules and airlocks be? At

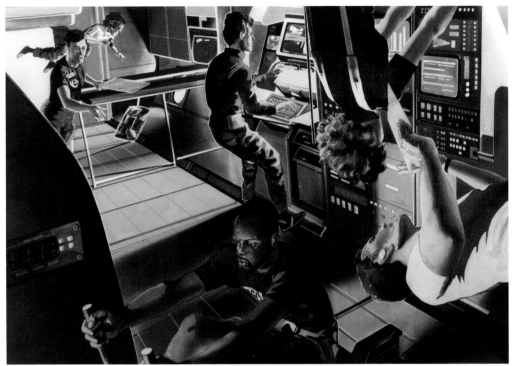

This artist's concept by Harold Smelser depicts what life might be like aboard Space Station Freedom. At bottom left, an astronaut exercises on a rowing machine while watching a video. Above him another astronaut jogs on a treadmill. Others work and prepare food for the crew. (NASA photo, no. 90-HC-550)

what air pressure should the station operate? What kind of fuel should the station use to re-boost itself? Where and how should that fuel be located and stored? With what method should the station generate the 75 kilowatts of electrical power required? How sophisticated should the environmental control and life support system be? What thermal system should be employed? Should structures be deployed or assembled in space, and if the latter, by what method? At what altitude should the station operate? How long should the periods between shuttle visits be? How many crew members per rotation?[28]

NASA engineers defined the number of American-built modules at four (two laboratory and two habitation), although this was later reduced. And the module length and mass kept increasing. In 1986 it stood at 44.5 feet. Designers established what they referred to as a figure-eight pattern for these modules, narrower in the middle than on the two ends. This would reduce the size of the module, and therefore its cost, while enabling more volume within each. The figure-eight

The Freedom program was supposed to be about cutting edge science in the microgravity environment of space. In this painting by Harold Smelser, crew members engage in various scientific activities in the U.S. laboratory. (NASA photo, no. 90-HC-549)

pattern of the module also appeared technically more attractive, because it gave NASA managers greater flexibility in terms of growth, assembly, and shuttle launches and for integrating non-U.S. modules into an international station. Designers set internal cabin pressure at 14.7 psia (pounds per square inch of atmosphere), with an oxygen partial pressure of 20 percent.[29]

Even as the Power Tower took shape, it metamorphosed into a new, larger configuration known as the Dual Keel because of a twin truss superstructure. Those in favor of the Dual Keel thought it would offer a superior facility for users, whereas the Power Tower configuration could not satisfy "high customer interest in increasing module/payload access opportunity to Space Station regions with low micro-G." Marshall Space Flight Center engineers championed the Dual Keel and identified advantages for a space station with two keels instead of one, an upper and a lower boom, creating more attachment space and moving modules to the middle of the station instead of at the bottom.[30] In spite of the possible extra costs that a Dual Keel configuration might pose, Freedom project manager John Hodge approved this fundamental technical alteration. At the same time he told his design team, "Budgetary criteria should continue to be a viable part of the decision-making process."[31]

Other technical changes followed. Each drove up the cost of the design, complicated the process, and increased the time needed to complete the system. For instance, NASA managers decided to change the environmental control and life support to a "distributed" system. That meant that every module would have its own subsystems, instead of a centralized system where critical life sustaining equipment would be located in only one or two modules or in some other place on the station. They also changed it to a "closed loop" (complete recovery of water) rather than an "open loop" system (little or no recovery of water). This would enable astronauts to recover, reuse, and recycle water through regenerative systems technology. In every case these changes violated attempts to shave front-end costs as much as possible, but the engineers justified them on the basis of enhancing long-term capability. These major configuration changes are synopsized in the appendix.[32]

Trapped in a cycle of changing design elements and alterations, a situation compounded by the controversial way Beggs originally had gained approval, the space station's political support proved shallow. This was the second major factor in Freedom's demise. The station represented, in essence, incremental space policy without a broad consensus that it was even necessary to accomplish some perceived critical need of the nation.[33] Because of the two-year life span of any given Congress, coupled with an annual budget process, politicians necessarily focus on short-term concerns. In so doing, they underscore the reality of the incremental nature of the political process. Two years may seem like sufficient time, especially with Congress remaining in session much of the time, but in reality it is very limiting.[34] As estimates of costs grew throughout the rest of the 1980s, NASA and Congress struggled to contain them. Redesign followed redesign. Automated platforms were deleted from the plan, and the occupied base was scaled back. Cost growth drove the redesigns, and then the 1986 Space Shuttle Challenger tragedy caused NASA to completely reassess the design. Among the changes was a decision to build a separate "lifeboat" capability from the Space Shuttle to ensure the astronauts could return to Earth in an emergency. With every change in the technical requirements, Congress became increasingly involved and pressure mounted to either complete the effort or cancel it.

That Congress did not terminate the program was in part because of the desperate economic situation in the aerospace industry—the result of a recession and of military demobilization after the collapse of the Soviet Union and the end of the Cold War—and the fact that by 1992, the project had spawned an estimated 75,000 jobs in thirty-nine states, most of which were key states such as California, Alabama, Texas, and Maryland. Politicians were hesitant to kill the station outright because of these jobs, but neither were they willing to fund it at the level required to make it truly viable. Senator Barbara Mikulski (Democrat, Maryland), chair of the Senate Appropriations subcommittee that handled NASA's budget, summarized this position: "I truly believe that in space station Freedom we are going to generate jobs today and jobs tomorrow—jobs today in terms of the actual manufacturing of space station Freedom, but jobs tomorrow because of what we will learn."[35]

Nevertheless, at a fundamental level the political process of government appropriations came to dominate the space station program. Remarkably, the program survived this incremental policy gauntlet throughout the 1980s, and of course its successor did so through the end of the twentieth century. Virtually any reading of the history of the Apollo program reveals that within two years—about the time the engineers actually began bending metal—the political consensus started unraveling. The program was sustained for the next several years in precisely the incremental way that was required by the American political process for anything of that nature.[36]

Even more remarkable was the fact that the space station was able to survive politically despite its not being viewed as critical to national security, as Apollo had been, despite waning support from the scientific community, despite lack of enthusiasm on the part of the general public, and despite rising budgetary deficits. In the end, between 1984 and 1992 Congress steadfastly supported the program—albeit sometimes with very close vote margins. The endurance of the program should be attributed to the adeptness of NASA at marshaling support when required and maintaining a coalition of intensely interested parties that ensured their positions were heard on Capitol Hill and in the White House.[37]

The third major factor in the development of Space Station Freedom was its cost, to which every technical and political decision seemingly added. High-technology programs routinely cost much more than originally projected. Some, such as clearly understood national defense programs, do not suffer. Most others do. Every time a new cost projection came out, NASA had to rally supporters to defend the program. They always did so based on some future fantastic capability, in the process promising more and more return on investment. This led to a disastrous spiral of overselling and coming up short, even as costs grew.

Of course, when NASA began the space station program in 1984, its assembly was to have been completed by 1994 at a cost of about $8 billion. By the time NASA terminated the Freedom program in 1993, that date had stretched to 2000. The estimates for the original design in 1984 did not include costs for launching the station into orbit, civil service salaries, or operational expenses. It was the cost growth of the space station during the latter 1980s that proved its undoing. The inability to contain cost led to several restructurings, and eventual cancellation, of Space Station Freedom.[38]

In January 1987, NASA's public admission that the station's development cost estimates had inflated by more than 80 percent since 1984, a proposed increase from $8 to $14.5 billion in fis-

This illustration emphasizes the pressurized modules where crews were to work and live. Four units, two provided by the United States and one each by Japan and Europe, are attached to a horizontal transverse boom. Resource nodes—which house the distributed subsystems as well as command and control stations—connect the laboratory and habitation modules. Two crew members inside the cupola atop the node on the right control the Canadian-provided Remote Manipulator arm. In the background, a co-orbiting platform flies in tandem with Freedom. An orbital maneuvering vehicle flies toward the platform with which it will rendezvous and, after attaching itself, brings the platform back to the station for servicing. At the bottom of the image, a Space Shuttle prepares to dock with the station. (NASA photo, no. 86-H-331)

cal year 1984 dollars, sparked outcries from every sector of government and among interest groups claiming that the money would be better spent elsewhere.[39] But this was not the end. In April another "scrub" of the program raised the total amount to $16 billion, and in 1989 another review estimated the cost at a whopping $30 billion.[40] NASA's inability to control this growth, even before anything had been built and launched into orbit, raised questions about the ability of the organization to carry out large scale technological enterprises. Seemingly, it had sunk low from its glory days of Apollo, when nothing seemed impossible. Coupled with the Challenger accident, other apparent failures, and weak management at the space agency, this cost growth had an enormous ripple affect from Washington to each of the NASA centers. Could NASA be trusted as it once had been? The answer for many was no. And for every member of Congress and leaders at the White House reaching that conclusion, intrusion into the management structure and questions about the program increased. The concerns brought more investigation and checking, and the outcome was predictable. The program wallowed in a bureaucratic morass, like a woolly mammoth trapped in the La Brea tar pits 40,000 years ago, unable to extricate itself. With every exertion it moved closer to death.

The fourth factor that fundamentally shaped the Freedom program was the international character that it had had from the outset. Although a range of cooperative activities had been carried out in the past—Spacelab, the Apollo-Soyuz Test Project, and scientific data exchange—the station offered an opportunity for a truly integrated effort. The inclusion of international partners, many with their own rapidly developing spaceflight capabilities, could enhance the effort. In addition, every partnership brought greater legitimacy to the overall program and might help to insulate it from drastic budgetary and political changes. Inciting an international incident because of a change to the space station was something neither U.S. diplomats nor politicians relished, and that fact, it was thought, could help stabilize funding, schedule, or other factors that might otherwise be changed in response to short-term political needs.[41]

NASA leaders understood these positive factors, but they also recognized that international partners would dilute their authority to execute the program as they saw fit. Throughout its history, the space agency had never been very willing to deal with partners, either domestic or international, as equals. It tended to see them more as a hindrance than help, especially when they might get in the way of the "critical path" toward any technological goal. Assigning an essentially equal partner responsibility for the development of a critical subsystem meant giving up the power to make changes, to dictate solutions, and to control schedules and other factors. *Partnership,* furthermore, was not a synonym for "contractor management," something agency leaders understood well, and NASA was not accepting of full partners unless they were essentially silent or at least deferential. Such an attitude militated against significant international cooperation.

In addition to this concern, some technologists expressed fear that bringing Europeans into the project really meant giving foreign nations technical knowledge that only the United States held. No other nation could build a space station on a par with Freedom, and only a handful had a genuine launch capability. Many government officials questioned the advisability of reducing the United States' technological lead. The control of technology transfer in the international arena was an especially important issue.[42]

In spite of these concerns, NASA leaders pressed forward with international agreements between thirteen nations to take part in the Space Station Freedom program. Japan, Canada, and the countries pooling their resources in the European Space Agency (ESA) agreed in spring 1985 to participate. Canada, for instance, decided to build a remote servicing system. Building on knowledge gained with its Spacelab science module for the Shuttle, ESA agreed to build an attached pressurized science module and an astronaut-tended free-flyer. Japan's contribution was the development and operation of an experiment module for materials processing, life sciences, and technology. These separate components, with their "plug-in" capacity, eased somewhat the management (and Congressional) concern about unwanted technology transfer.[43]

The fifth major issue, the space station's suffering from lack of support from the scientific community for whom it was potentially being built, had caused controversy almost from the outset. Many scientists believed the money for the station could be much more effectively spent on other science. They viewed it as a zero sum game; if the funds went to Freedom, they would not be available elsewhere. If Freedom were canceled, these same people thought that their specific programs could then be funded. But as NASA pared away at the station budget, it also eliminated functions that some of its constituencies wanted. This led to a rebellion among some former supporters. For instance, the space science community began complaining that the configuration under development did not provide sufficient experimental opportunity. Thomas M. Donahue, an atmospheric scientist from the University of Michigan and chair of the National Academy of Science's Space Science Board, commented in the mid-1980s that his group saw "no scientific need for this space station during the next twenty years." He also suggested, "if the decision to build a space station is political and social, we have no problem with that," alluding to the thousands of jobs associated with it. "But don't call it a scientific program."[44]

THE DEMISE OF THE SPACE STATION FREEDOM PROGRAM

Unable to make positive strides toward completion, redesigns of Space Station Freedom took place in 1990, 1991, 1992, and 1993. The results were devastating, as the station declined in capability each time. One attempt at gallows humor at NASA in the midst of these redesigns

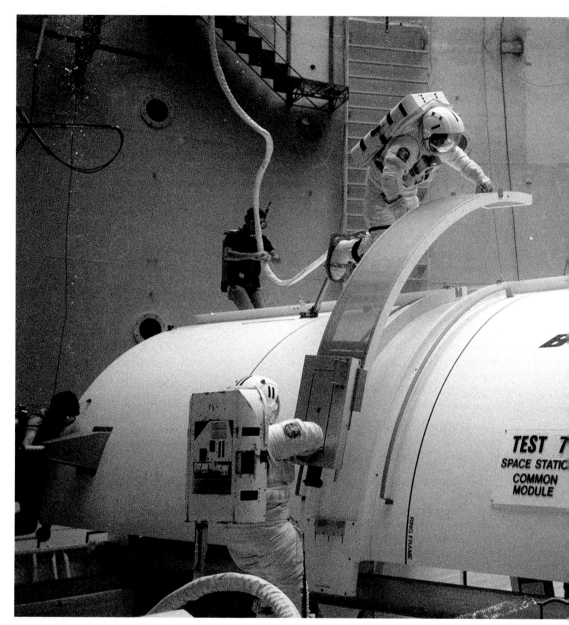

This image shows two astronauts practicing construction techniques to build a space station in the Neutral Buoyancy Simulator (NBS) at Marshall Space Flight Center in 1985, early in the space station program. The NBS tank is 75 feet in diameter and 40 feet deep, and it contains 1.3 million gallons of water. (NASA photo, no. 8562087)

demonstrates the point: it shows a drawing of a NASA project manager sitting at a desk with a small space station model on it. A colleague stands before him and says, "Hey Joe, great model of Freedom!" Joe responds, "What do you mean *model?*" The program had become essentially worthless as it could no longer perform many of the functions for which it had been conceived. Even so, NASA kept the project together, mostly for bureaucratic reasons. If the agency did not have a large project to complete, how could it justify its size and cost? NASA could come forward with a more viable effort at a later date. It chose to fight for Freedom.[45]

Instead, in the fall and winter of 1992–93 the Space Station Freedom program was transformed into a new design that NASA said would use 75 percent of Freedom's hardware and systems. Russia was added as another international partner at the same time. The program was renamed International Space Station Alpha, and, later, simply International Space Station. There would be two U.S., one European, one Japanese, and five Russian modules (three for science), with habitation for a crew of six. Canada would build a Mobile Servicing System. The station would be located in 51.6° orbit (to allow access from Russia), the operating period shortened from thirty to ten years, and annual operating costs reduced.[46]

In 1984 opposition to the Soviet Union had been the driving force behind the approval to move forward with what became Space Station Freedom. Ironically, what probably best explains the reason that any space station program continued to exist in the 1990s was an alliance intended to prop up the fledgling Russian democracy in the aftermath of the Cold War. This would, some politicians believed, prevent Russia from backsliding into an evil empire, and it would deter nuclear proliferation.

The Freedom program yielded nothing in orbit, though some of the hardware and much of the research and development would later find use on the International Space Station. It also played an important social function. Throughout history, people have performed tasks and constructed wonders that have had little to do with profit or the needs of the marketplace, and everything to do with faith, vision, harmony, and prestige. A space station as a symbol of humankind's planetary unity may be the latest example of this drive, aside from its serving as an icon for the advancement of science and technology. However, in the end NASA's leaders did not prove a match for the challenge.

The Space Station Freedom program also had more practical applications in domestic politics. One of the reasons why it was stuck in a redesign cycle from 1987 to 1993 was that its existence served to employ thousands of highly skilled technical people, in both the private and public sectors. These individuals, and the tax money supporting the effort, were scattered throughout Congressional districts around the nation. For instance, Senator Barbara Mikulski said the program was about "jobs, jobs, jobs."[47] Seen in this harsh light, a space station program of any incarnation is a works project for the technological elite, and it therefore makes as much sense politically as any other of the many government programs designed to redistribute wealth across the nation. It also explains why a redesign cycle is not only tolerated but might prove necessary: success ends the program.

But there were sound international political reasons to start and maintain Space Station Freedom. U.S. policy in the 1980s prevented unchecked competition by unifying all non–Soviet Union space programs under one banner, even as it preserved national technological advantage with a serious program to restrict technology transfer. NASA managers and State Department officials made sure that access to information on the results of experiments performed aboard a space station would be proprietarily restricted. Thus, station supporters could advocate the program to Congress on the simultaneous grounds of creating international cooperation while also preserving the economic advantage of the United States.

Finally, there is the technical result. NASA and the overall aerospace community learned much in the process of designing Space Station Freedom, even if the hardware was not completed. The station, once built, would also serve as a proving ground for spaceflight technologies in Earth orbit. All spaceflight requirements tended to point toward using the station as a means to test materials and technologies in a radiation-filled, microdebris-strewn, microgravity environment and as a command/control/communications post for future activities. This could even have a military purpose, which spurred the Soviet Union to move forward with its own space station, Mir, in 1986. There was no easy way to determine in advance whether the Soviet Union would ever come up with any information from Mir that could be used to produce an operational space-based weapons system or improve a ground-based one. The hard way of learning, however, was a scenario that national security interests dreaded most: a technological surprise, a sudden vulnerability.

As a space station program official once commented, the design grew gracefully, but it did not shrink gracefully. Budget cuts became commonplace, but without firm priorities program managers could not cut out something completely, because they did not have the political authority to do so. Instead they cut corners, increased the overall technological risk, and redesigned in a penny-wise fashion. Managers came and went, soon no one had a good idea of the overall point of the program, and later the only meaning that remained was that somehow it had to go on for another year.[48]

The lesson about Space Station Freedom that may be the most difficult to accept is the possibility that democracies, except in times of real or perceived crisis, are virtually unable to establish and maintain large-scale scientific and technological priorities. It was a fairly simple undertaking for dictators, emperors, pharaohs, and kings to dictate the plans and means for impressive public monuments. But a modern democratic republic such as the United States has trouble with similar complex tasks. The dictates of the marketplace, not the White House and Congress, are supposed to determine what kinds of innovations will occur. Space Station Freedom was an example of heterogeneous engineering, which recognizes that technological issues are simultaneously organizational, economic, and political. A complex web of ties between various people, institutions, and interests brought forward a Freedom program that could never be built.[49] It finally died after years of agony. That the International Space Station rose from the ashes of Freedom, and met with some success, informs the remainder of this study.

141

THE STRANGE

CAREER OF

SPACE STATION

FREEDOM

CHAPTER 5

A Mir Interlude

E ven as the United States' human spaceflight effort was paralyzed beginning January 28, 1986, in the wake of the Space Shuttle Challenger accident, the Soviet Union launched its Mir station on February 20, 1986. The latest in a succession of increasingly sophisticated and capable orbital workshops, Mir remained operational in orbit until March 2001. It was the site of many important activities for the Soviet Union, among them in 1987 and 1988 when several cosmonauts set records for duration in Earth orbit.

Mir also became the centerpiece, in 1992, of an international program in which it was expected to become an early step in developing the International Space Station. In 1995 the

Russia's Mir space station was launched in 1986 and served an important purpose until finally abandoned in 1998 and deorbited in 2001. Here it is photographed from the Space Shuttle Atlantis during its first docking mission with Mir, STS-71. (NASA photo, no. 95-HC-401)

United States and Russia began the Shuttle-Mir program, in which the Space Shuttle docked with Mir in a succession of nine missions through 1998, and American astronauts undertook extended stays on the Russian craft. There were several problems. During the attempted docking of the Russian resupply vessel Progress with Mir on June 25, 1997, the vessel collided with the science module Spektr, which was attached to the station. The module decompressed, and its solar arrays were knocked out of service. Although the crew of two Russian cosmonauts

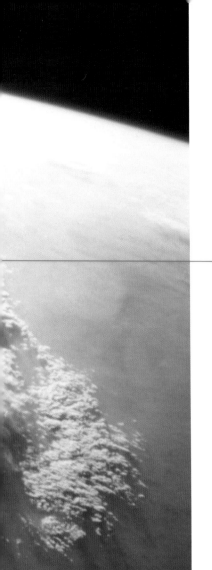

As seen from Atlantis, this shot of Mir was taken during a docking mission on January 14, 1997. (NASA photo, no. 97-HC-114)

and one American astronaut were uninjured, the accident crippled the station and led to a series of crises in space. In a courageous decision, the Russian Space Agency refused to abandon the station and managed to keep it operational until it could be resupplied, with its critical systems restored. Despite political pressure to remove American astronauts from Mir after the accident, NASA officials assessed the risks and decided to continue the cooperative missions. Mir remained in orbit with a crew until 1999, when the Russians prepared it for deorbit. There were numerous reconsiderations over this decision, but in March 2001 it finally returned to Earth, impacting into the South Pacific.

MIR MODULE DESCRIPTIONS

The Mir core resembled Salyut 7 but had six ports instead of two. Fore and aft ports were used primarily for docking. Four radial ports in a node at the station's front were for berthing large modules. The core weighed 20.4 tons at launch in 1986.

Kvant was added to the Mir core's aft port in 1987. This small, 11-ton module contained astrophysics instruments and life support and attitude control equipment.

Kvant 2, added in 1989, carried an EVA airlock, solar arrays, and life support equipment. The 19.6-ton module was based on the transport logistics spacecraft originally intended for the Almaz military space station program of the early 1970s.

Kristall, added in 1990, carried scientific equipment, retractable solar arrays, and a docking node equipped with a special androgynous docking mechanism designed to receive spacecraft weighing up to 100 tons. Originally the Russian Buran shuttle, which made one unmanned orbital test flight in 1988, would have docked with Mir using the androgynous unit. Space Shuttle Atlantis used the androgynous unit to dock with Mir for the first time on the STS-71 mission in July 1995. On STS-74, in November 1995, Atlantis permanently attached a Docking Module to Kristall's androgynous docking unit. The Docking Module improved clearance between Atlantis and Mir's solar arrays on subsequent docking flights. The 19.6-ton Kristall module was based on the transport logistics spacecraft originally designed to carry Soviet soldier-cosmonauts to the Almaz military space stations.

Spektr was launched on a Russian Proton rocket from the Baikonur launch center in central Asia on May 20, 1995. The module was berthed at the radial port opposite Kvant 2 after Kristall was moved out of the way. Spektr carried four solar arrays and scientific equipment, including more than 1600 pounds of U.S. equipment. The focus of scientific study for this module was Earth observation, specifically natural resources and atmosphere. The equipment onboard was supplied by both Russia and the United States.

Priroda was the last science module to be added to the Mir. Launched from Baikonur on April 23, 1996, it docked to the space station as scheduled on April 26. Its primary purpose was to add Earth remote sensing capability to Mir. It also contained the hardware and supplies for several joint U.S.-Russian science experiments.

The Docking Module was delivered and installed by Shuttle mission STS-74 in November 1995, making it possible for the Space Shuttle to more easily dock with Mir. On STS-71 in June 1995, the Shuttle docked with the Kristall module on Mir. However, to make that docking possible, the Kristall configuration had to be changed to give the Shuttle enough clearance to dock. Russian cosmonauts performed a spacewalk to move the Kristall module from a radial axis to a longitudinal axis, relative to Mir. After the Shuttle departed, Kristall was moved back to its original location.

Source: NASA Johnson Space Center Fact Sheet, "International Space Station—Russian Space Stations, June 1997," NASA Historical Reference Collection.

LAUNCHING MIR

Mir's launch in 1986 represented the penultimate in Soviet space technology. Weighing 20.4 tons at launch, with 3,000 cubic feet of habitable space, Mir was the core of a projected larger facility. By the early 1990s, the complex had grown to a weight of about 110 tons—with 13,000 cubic feet of habitable space—because of additional modules. At the time of the restructuring of the Soviet Union, the station consisted of the Mir core and the modules Kvant 1 (launched March 31, 1987), Kvant 2 (launched November 26, 1989), and Kristall (launched May 31, 1990). By that time Mir was more than 107 feet long, with its docked Progress-M resupply and Soyuz-TM crew transfer spacecraft, and about 90 feet across at its modules. Although the core resembled Salyut 7, its immediate predecessor, it had six ports with those fore and aft used primarily for docking.[1]

The Soviet Union's purpose for the Mir program, like that of the U.S. Freedom space station, emphasized a series of scientific and technical activities. One of the most important was an attempt to maintain a permanent Soviet presence in space and to learn about the requirements necessary for long-duration human spaceflight. In that regard, the Mir program proved enormously fruitful. Except for two periods—July 17, 1986, to February 4, 1987, and April 27 to September 4, 1989—cosmonauts occupied Mir continuously. The longest-duration Soviet mission took place between December 21, 1987, and December 21, 1988: a total of 366 days. On that mission, cosmonauts Vladimir Titov and Musa Manarov far outdistanced any previous spacefarers after a period of one year, twenty-two hours, and thirty-nine minutes in space. They even broke their colleague's record; Yuri Romanenko had established a record of 326 days in Earth orbit between February 5 and December 29, 1987. Even these long-duration missions paled in comparison to an incredible 439 days in orbit for Valery Polyakov between January 8, 1994, and March 22, 1995, leading to a set of cooperative missions with the United States aboard Mir starting in the summer of 1995.[2]

From March 1986 to December 1992, the Mir complex supported twelve long-duration missions, most four to six months and the longest lasting one year. Nine international crew members visited Mir on shorter flights during the same period. The upgraded Soyuz-TM human transport spacecraft entered regular service to support Mir, and about thirty robotic Progress cargo craft were used to resupply and refuel the station. Numerous tasks—maintenance, repair, construction—and new-technology demonstrations were also carried out during thirty-one two-person extravehicular activities (EVAs) totaling about 140 hours prior to the U.S.-Russian cooperative program that began in 1995.[3]

The Soyuz spacecraft used to ferry cosmonauts to and from Earth also found service in occasional "fly-around" missions orbiting the T-shaped Mir. During the entire range of Mir's life, including the cooperative era with the United States, crew members spent more than 325 hours as part of seventy-five EVAs to conduct research and repairs on the station. More than half of these came during the last six years of Mir's existence. Additional hours were spent during three intravehicular walks inside an unpressurized Spektr module. Participants in the EVAs included twenty-nine Russian cosmonauts, three U.S. astronauts, two French astronauts, and one European Space Agency astronaut, a citizen of Germany. Cosmonaut Anatoly Solovyev donned the Russian Orlan space suit for sixteen spacewalks for a total time of seventy-seven hours, forty-six minutes: more EVA time than any other spacewalker in the world.[4]

At a fundamental level, Mir should be considered a tremendous success story for the Soviet Union and a representation of both that nation's past space glories and its future as a leader in space. Mir endured fifteen years in orbit, three times its planned lifetime. It outlasted the Soviet Union that launched it into space, hosting scores of crew members and international visitors. It raised the first crop of wheat to be grown from seed to seed in outer space. Mir was the scene of joyous reunions, feats of courage, moments of panic, and months of grim deter-

During the first Shuttle rendezvous mission with Mir, February 8, 1995, Russian cosmonaut Valeriy V. Polyakov, who would be aboard the station for more than a year, looks out the window at Discovery. (NASA photo, no. STS063-711-080)

Astronaut Robert "Hoot" Gibson *(right)*, STS-71 mission commander, shakes the hand of his Russian counterpart, Mir-18 commander Vladimir N. Dezhurov *(left)*, June 29, 1995. This was the first Shuttle docking mission to Mir. (NASA photo, no. 95-HC-444)

mination. It suffered dangerous fires, a nearly catastrophic collision, and desperate periods of out-of-control tumbling.[5] Traveling at an average speed of 17,885 mph, the station orbited about 250 miles above Earth and could be seen on clear nights as a bright light racing across the northern sky.

Always controversial, Mir was both a noble and a foolhardy venture. It represented the best and the worst of Soviet technology and society. It was robust, coarse, accident-prone, and a marvel. Some called it a lemon, which indeed it was some of the time, but it was also an oasis, preserving life in the treacherous "desert" of space. Although the name Mir has often been rendered into English as "peace," no single-word translation captures its significance. It symbolized, perhaps almost as effectively as the Red Army's military capability, the superpower status of the Soviet Union. It also provoked reflections on what the human race is and on its future destiny. Some have described Mir as a giant Tinker Toy, a term that recalls its construction and ungainly physical characteristics. Adding modules over the years and then sometimes rearranging them, the Russians assembled the strangest, largest structure ever placed in Earth orbit. One may appropriately extend to it Smithsonian National Air and Space Museum archivist Brian Nicklas's criteria for identifying aircraft:

> *If it's ugly, it's British.*
> *If it's weird, it's French.*
> *If it's ugly and weird, it's Russian.*[6]

Mir combined the bluster of former Soviet Union premier Nikita Khrushchev, the grace of Russian-American ballet dancer Mikhail Baryshnikov, the genius of Russian author and historian Aleksandr Solzhenitsyn, the paranoia of former Soviet Union leader Joseph Stalin, and the brilliance of Russian nuclear physicist Andrei Sakharov to create a weird, ugly, and highly successful space vehicle.

In outward appearance, Mir has been compared to a dragonfly with wings outstretched, another appropriate physical characterization.[7] NASA astronaut Jerry Linenger, who flew on Mir in the mid-1990s, compared the station to

> six school buses all hooked together. It was as if four of the buses were driven into a four-way intersection at the same time. They collided and became attached. All at right angles to each other, these four buses made up the four Mir science modules. . . . Priroda and Spektr were relatively new additions. . . . and looked it—each sporting shiny gold foil, bleached-white solar blankets, and unmarred thruster pods. Kvant 2 and Kristall . . . showed their age. Solar blankets were yellowed. . . . and looked as drab as a Moscow winter and were pockmarked with raggedy holes, the result of losing battles with micrometeorite and debris strikes over the years.[8]

Inside Mir looked a cluttered mess, with obsolete equipment, floating bags of trash, the residue of dust, and a crust that grew more extensive with the passing years. Astronaut Michael Foale said it reminded him of "a frat house, but more organized and better looked after."[9]

THE BEGINNINGS OF U.S.-RUSSIAN COOPERATION ON MIR

During the same time period as the restructuring of the Soviet Union in 1991, NASA opened negotiations to undertake cooperation with the new Russian Space Agency to extend knowledge about long-duration missions beyond what was possible with the Space Shuttle. Cooperation with the Soviet Union had previously been highly limited, with the most important projects being undertaken only as an attempt to open lines of political communication between the two superpowers.[10] On July 1, 1991, the Soviet Union's military alliance, the Warsaw Pact, dissolved as numerous satellite countries withdrew from it. This signaled perhaps better than any other single event the collapse of the Soviet Union. Almost immediately Oleg Shishkin, minister of General Machine Building in the former Soviet Union, met with U.S. vice president Dan Quayle to discuss a venture in which the United States and Russia would cooperatively use Mir for human missions. They quickly concluded deliberations, and on July 31, 1991, President George Bush and Premier Mikhail Gorbachev signed an agreement whereby an American astronaut would reside on Mir for up to six months performing biomedical experiments, and a Russian cosmonaut would fly on the Space Shuttle. The agreement also established a Manned Flight Joint Working Group to coordinate these activities.[11]

In April 1992 the Russian Federation, principal successor to the Soviet Union, created a civilian space organization, the Russian Space Agency. Its leader was Yuri Koptev, formerly an official of the Soviet Ministry of General Machine Building. On April 1, 1992, new NASA administrator Daniel S. Goldin took office. The two agency heads met for the first time in June 1992 and quickly agreed that there were many opportunities for enhanced cooperation, particularly in the area of human spaceflight. During a summit meeting between Russian president Boris Yeltsin and U.S. president George Bush a few days later, the two countries announced their intention to broaden cooperative relations in space. On June 17, 1992, they issued a "Joint Statement on Cooperation in Space." There were two key provisions, according to the White House press release: flights of Russian cosmonauts aboard a Space Shuttle mission (STS-60) and U.S. astronauts aboard the Mir space station in 1993; and a rendezvous and docking mission between Mir and the Space Shuttle in 1994 or 1995.[12] Cooperation between the two nations in space accelerated under the next U.S. president, Bill Clinton.

As a result of the U.S.-Russian dialogue on expanded space cooperation initiated in June 1992, the Russian Space Agency and NASA signed an agreement in October 1992 to exchange cosmonauts and astronauts on each other's human spaceflight missions. It also established the protocols for docking the Space Shuttle with the Mir space station. During its first year in office, accordingly, the Clinton administration moved to substantially expand existing U.S.-Russian cooperation, in effect merging large portions of the efforts of the only two countries with the capability of sending people into space; such a move was announced in September 1993. The political decision to undertake this expansion was linked to broader foreign policy concerns, such as stemming the proliferation of missile technology capability and providing job opportunities for the Russian aerospace sector.[13]

Russian cosmonaut Gennadiy M. Strekalov engages in one of five spacewalks conducted by the Mir-18 crew between March and July 1995. (NASA photo, no. 95-HC-459)

After a few more months of discussion, NASA and the Russian Space Agency announced a comprehensive cooperative program to jointly use Mir. They determined that a decade-long program of human spaceflight would involve three distinct phases. Phase One (1994–97) was fundamentally an expansion of the Human Spaceflight Agreement of October 1992, to include seven to ten Shuttle flights to Mir in addition to five medium- to long-duration flights on Mir by U.S. astronauts. Phase Two (1997–98) involved U.S., Russian, and Canadian elements. It achieved the ability to support three people in 1998, with the delivery of the Soyuz-TM crew-rescue vehicle to service an International Space Station then under construction. Phase Three (1998–2002) completed assembly of the ISS, including European and Japanese components, and emphasized station operations.[14]

On December 16, 1993, the heads of the two agencies signed a protocol reflecting this new level of activity. The following provisions are quoted from the U.S.-Russian agreement:

- An additional Russian cosmonaut flight on the Space Shuttle will take place in 1995. The back-up cosmonaut currently in training at NASA's Johnson Space Center will be the primary cosmonaut for that flight, with the STS-60 primary cosmonaut acting as back up. During this mission, the Shuttle will perform a rendezvous with the Mir-1 space station and will approach to a safe distance, as determined by the Flight Operations and Systems Integration Joint Working Group established pursuant to the October 5, 1992 Agreement.

- The Space Shuttle will rendezvous and dock with Mir-1 in October-November 1995, and, if necessary, the crew will include Russian cosmonauts. Mir-1 equipment, including power supply and life support system elements, will also be carried. The crew will return on the same Space Shuttle mission. This mission will include activities on Mir-1 and possible extravehicular activities to upgrade solar arrays. The extravehicular activities may involve astronauts of other international partners of the Parties.

- NASA-designated astronauts will fly on the Mir-1 space station for an additional 21 months for a Phase One total of two years. This will include at least four astronaut flights. Additional flights will be by mutual agreement.

- The Space Shuttle will dock with Mir-1 up to ten times. The Shuttle flights will be used for crew exchange, technological experiments, logistics or sample return. Some of those flights will be dedicated to resources and equipment necessary for life extension of Mir-1. For schedule adjustments of less than two weeks, both sides agree to attempt to accommodate such adjustments without impacting the overall schedule of flights. Schedule adjustments of greater than two weeks will be resolved on a case-by-case basis through consultations between NASA and RSA.

- A specific program of technological and scientific research, including utilization of the Mir-1 Spektr and Priroda modules, equipped with U.S. experiments, to undertake a wide-scale research program, will be developed by the Mission Science Joint Working Group established pursuant to the October 5, 1992 Agreement. The activities carried out in this program will expand ongoing research in biotechnology, materials sciences, biomedical sciences, Earth observations and technology.

- Technology and engineering demonstrations applicable to future space station activities will be defined. Potential areas include but are not limited to: automated rendezvous and docking, electrical power systems, life support, command and control, microgravity isolation system, and data management and collection. Joint crew operations will be examined as well.

- The Parties consider it reasonable to initiate in 1993 the joint development of a solar dynamic power system with a test flight on the Space Shuttle and Mir in 1996, the joint development of spacecraft environmental control and life support systems, and the joint development of a common space suit.
- The Parties will initiate a joint crew medical support program for the benefit of both sides' crewmembers, including the development of common standards, requirements, procedures, databases, and countermeasures. Supporting ground systems may also be jointly operated, including telemedicine links and other activities.
- The Space Shuttle will support the above activities, including launch and return transportation of hardware, material, and crewmembers. The Shuttle may also support extravehicular and other space activities.
- Consistent with U.S. law, and subject to the availability of appropriated funds, NASA will provide both compensation to the RSA for services to be provided during Phase One in the amount of US $100 million in FY 1994, and additional funding of US $300 million for compensation of Phase One and for mutually-agreed upon Phase Two activities will be provided through 1997. This funding will take place through subsequent NASA-RSA and/or through industry-to-industry arrangements. Reimbursable activities covered by the above arrangements and described in paragraphs 3 through 8 will proceed after these arrangements are in place and after this Protocol enters into force in accordance with Article III. Specific Phase One activities, schedules and financial plans will be included in separate documents.
- Implementation decisions on each part of this program will be based on the cost of each part of the program, relative benefits to each Party, and relationship to future space station activities of the Parties.[15]

Thus the United States agreed to provide funding for various Russian activities and hardware associated with the expanded cooperation, ending a longstanding NASA tradition that its cooperative programs must not involve an exchange of funds.

From the beginning, many in the United States questioned the desirability of undertaking a major cooperative effort with the new Russian nation. The reasons for this hesitancy rested on half a dozen key concerns, as quoted from a report by the U.S. Congress Office of Technology Assessment (OTA) in 1995:

Technical risks. Despite Russia's prowess in developing and maintaining a large and capable space program, it has certain weaknesses, such as difficulty maintaining schedules on

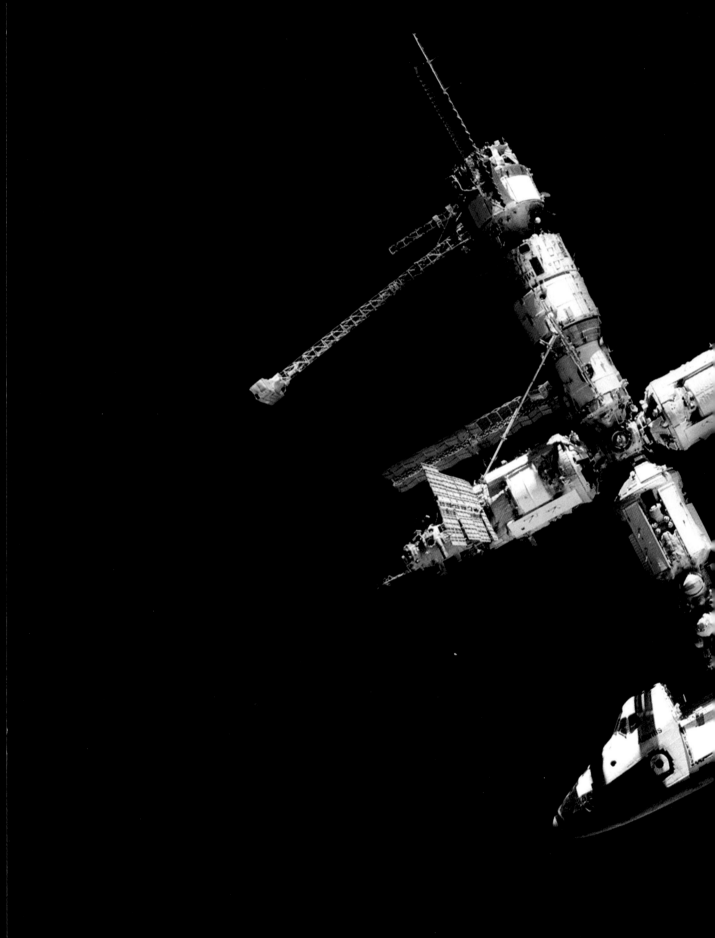

Cosmonauts Anatoliy Y. Solovyev and Nikolai N. Budarin took this photograph on July 4, 1995, from their Soyuz spacecraft during a fly-around of Mir. They enjoyed this view of the Space Shuttle Atlantis as it was docked to the Russian station. (NASA photo, no. STS071-S-072)

new spacecraft and components, which were evident even before the end of the Cold War. Russia will have to complete several new systems to fulfill its upcoming cooperative and contractual commitments.

Unstable political institutions. Russian democratic institutions are in a very early stage of development, and successful maturation is far from certain. Legal and political instability is great and appears likely to remain so for some time to come.

Russian military actions. The Russian military has undergone substantial change in the past few years and is much less stable than it was under the U.S.S.R. government. Instability in the Russian military could make the Western world much more wary about investing in Russia and could even undermine economic and political stability. For example, the war in Chechnya has drained important resources from the civilian economy and has raised concerns about human rights abuses.

Economic uncertainty. The near collapse of the Russian economy and its impact on the many enterprises essential to Russian space activity could affect Russia's ability to deliver on international commitments. Russia lacks a common, settled business and procedural framework within which to organize and regulate its new marketplace.

Crime and corruption. The political and legal changes in Russia and lax enforcement have increased the incidence of serious crime and open corruption, thus impeding the development of normal business relationships.

Cultural barriers. U.S. and Russian partners face a high risk of misunderstanding each other's intentions and of inadvertently creating discord in their relationships.[16]

Despite these concerns, the benefits of cooperating with the Russians on Mir far outweighed liabilities. There were four overarching geopolitical considerations that prompted the U.S.-Russian cooperative space program in the 1990s. First, it created a positive image of the United States in the international setting. Second, it encouraged greater public interaction between the United States and Russia. Third, it reinforced the perception of American openness to outside nations. And fourth, it expanded the use of space technology as a tool of diplomacy to serve broader U.S. foreign policy goals.[17] To these one might add five additional technical and economic reasons, as enumerated in an excerpt from a 1995 OTA report:

Reducing costs and sharing burdens. Many of the agencies involved in space share common goals and have developed overlapping programs. Facing budget constraints, these agencies are looking for ways to coordinate their programs to eliminate unnecessary duplication and to share the cost burden of projects they might otherwise do on their own.

Broadening sources of know-how and expertise. Scientists and engineers from other countries may possess technology or know-how that would improve the chance of project success.

Increasing effectiveness. The elimination of unnecessary duplication can also free up resources and allow individual agencies to match their resources more effectively with their plans. This reallocation of resources can eliminate gaps that would occur if agency programs were not coordinated. International discussions can be valuable even if they merely help to identify such gaps, but they can be particularly useful if they lead to a division of labor that reduces those gaps.

Aggregating resources for large projects. International cooperation can also provide the means to pay for new programs and projects that individual agencies cannot afford on their own. This has been the case in Europe, where the formation of the European Space Agency has allowed European countries to pursue much more ambitious and coherent programs than any of them could have accomplished alone.

Promoting foreign policy objectives. Cooperation in space also serves important foreign policy objectives, as exemplified by the International Space Station program. The agreements on space cooperation reached in 1993 and 1994 by Vice President Al Gore and Russian Prime Minister Viktor Chernomyrdin have also led to significant cooperative activities in space science and Earth observations.[18]

In addition, the Shuttle-Mir program provided NASA with opportunities, challenges, and lessons to learn in preparation for building the International Space Station. Four goals were identified to guide the program's development and management. These proved more than compelling and led directly to the first cosmonaut's flight on the Space Shuttle in 1994 and, all in 1995, the first Shuttle flight to Mir, the first American astronaut aboard the station, and the first docking mission of the Space Shuttle to Mir that July.

MISSION TO MIR

Accordingly, twenty years after the summer of 1975, when the world's two greatest spacefaring nations and Cold War rivals staged a dramatic docking in the Apollo-Soyuz Test Project, the space programs of the United States and Russia again met in Earth orbit. Phase One began in February 1994 with STS-60, when cosmonaut Sergei Krikalev worked beside American astronauts in Space Shuttle Discovery. It continued in February 1995, when Discovery rendezvoused with Mir during the STS-63 mission with cosmonaut Vladimir Titov aboard. On March 16, 1995, U.S. astronaut Norman E. Thagard, M.D., lifted off aboard the Russian Soyuz-TM 21 mission with two

Russian cosmonauts for a 115-day stay on Mir, one of the most significant missions to take place in recent years. Thagard's flight broke the 1973 Skylab 4 crew's U.S. record for eighty-four days in space. He and the two Russian members of the Mir-18 crew, Vladimir Dezhurov and Gennadiy Strekalov, were also the first Mir program crew to return to Earth via the Space Shuttle.[19]

Norman Thagard set the stage for all that followed: he was the first American astronaut to train in Russia, the first to launch aboard a Soyuz (with cosmonauts Vladimir Dezhurov and Gennady Strekalov), and the first to complete a residency aboard Mir. His training in Russia pioneered the pattern for the future. His scientific and medical investigations both increased knowledge in those areas and prepared the way for other American Mir astronauts. In his oral history of his flight, Thagard said: "I thought it was extremely ironic, because when I was flying missions in Vietnam in 1969 as an F-4 pilot, I thought that there was an excellent chance that at some point in time I'd have interactions with the Russians, but I thought it would be of a somewhat different nature than they turned out. If anyone in 1969 had ever told me that I would wind up having a captain in the Russian force as a commander, I would have said, 'You're crazy.'"[20]

Thagard also came away from the mission with a much greater understanding of the challenges of long-duration international spaceflight:

The nature of the flight's different. Shuttle flights are short, so you can intensively train for virtually every aspect of them, and that's not true for a three-month flight. . . . In fact, you're going to [have] things happen during the course of the flight that you never anticipated at all. It's possible in the future, with even longer flights, that there will be activities, experiments, space walks that were never foreseen at the time that you trained. . . . I think that aspect is one we're going to have to get used to, with the International Space Station. It's decidedly different than the way you approach a Shuttle Program.[21]

Thagard had some other difficulties. He suffered from space sickness (mostly a stomach ailment similar to motion sickness, but compounded by weightlessness) aboard Mir, and it was sometimes difficult to get his research done with pressures to carry out other duties. Also, he was aware of an ever-present set of cultural differences that divided him from the other the crew members.

After undertaking a set of biomedical experiments, Thagard returned home on the Space Shuttle Atlantis, STS-71, when it docked with Mir in July 1995. "This flight heralds a new era of friendship and cooperation between our two countries," said NASA administrator Daniel S. Goldin at the time. "It will lay the foundation for construction of an international Space Station later this decade."

This mission by Atlantis was the first of nine Shuttle-Mir linkups between 1995 and 1998,

Mir was a messy place to live and work. Here, for example, is the interior of the Spektr module in 1995. Cosmonaut Vladimir Dezhurov reroutes cable as part of the module's activation process. (NASA photo, no. 95-HC-454)

including rendezvous, docking, and crew transfers. Robert L. "Hoot" Gibson commanded the STS-71 crew, and his pilot was Charlie Precourt. Gibson was no novice to spaceflight, having made four previous flights, including command of the STS-41-B, STS-61-C, and STS-47 missions. The three STS-71 mission specialists aboard Atlantis—Ellen S. Baker, Bonnie Dunbar, and Gregory J. Harbaugh—were also veterans of spaceflight, having undertaken ten missions among them. Also aboard were cosmonauts Anatoly Y. Solovyev, making his fourth spaceflight, and Nikolai M. Budarin, making his first flight. Solovyev and Budarin were designated as the Mir-19 crew and would remain aboard Mir when Atlantis undocked from the nine-year-old space station and returned to Earth with the Mir-18 crew.22

Atlantis lifted off on June 27, 1995, from the Kennedy Space Center's Launch Complex 39-A in Florida. It was a significant effort to meet the narrow launch time that enabled the Shuttle to rendezvous with Mir. The available period, or "window," was approximately five minutes each day. Atlantis's docking with Mir actually began with the precisely timed launch and setting a

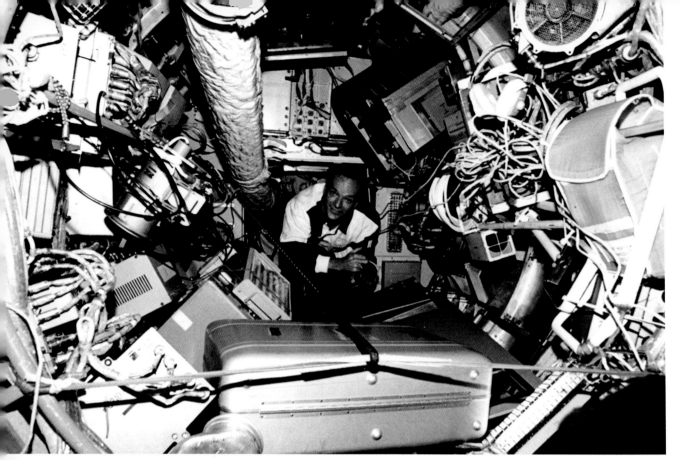

Astronaut Charles J. Precourt, STS-71 pilot, floats from Atlantis into the Kristall module of Mir during a historic eleven-day Shuttle docking mission that involved a total of ten astronauts and cosmonauts. (NASA photo, no. 95-HC-415)

course. Over the following two days, periodic firings of Atlantis's small thruster engines gradually brought the Shuttle closer to Mir.

Unlike most rendezvous procedures that typically have the Shuttle approaching from directly in front of its target, Atlantis aimed for a point directly below Mir along the Earth radius vector (R-Bar), an imaginary line drawn between the space station's center of gravity and the center of Earth. Approaching along the R-Bar allowed natural forces to brake Atlantis more than would have occurred following a standard Shuttle procedure. The R-Bar technique reduced the small number of jet firings close to Mir, thereby avoiding damage or contamination of its electricity-producing solar panels.

When Atlantis docked with Mir on July 29, it was perhaps the most significant event in the history of spaceflight since the symbolic joining of Apollo and Soyuz spacecraft twenty years earlier. It signaled a new age of cooperation in space, where exploration of the universe would be measured more by the achievements of coalitions than of single states.

After the ceremonies following the docking, the two groups of spacefarers undertook several days of joint scientific investigations inside the Spacelab science module tucked in Atlantis's large cargo bay. Research in seven different medical and scientific disciplines begun during Mir-18 concluded on STS-71. All of the experiments took advantage of the unique microgravity environment present on the spacecraft. Researchers used this capability to enhance human knowledge about spaceflight, especially the physiological changes of the body in weightlessness, and to understand anemia, high blood pressure, osteoporosis, kidney stones, balance disorders, and immune deficiencies. Thus the science conducted could hold consequences for humanity's survival beyond Earth as well as having significant terrestrial applications.

At the end of joint docked activities on July 4, 1995, Solovyev and Budarin assumed responsibility for operations of the Mir station. At the same time, the Mir-18 crew, who had been aboard since 16 March 1995—Commander Vladimir Dezhurov, flight engineer Gennady Strekalov, and American astronaut Norm Thagard—joined the STS-71 crew for the return trip to Earth.

After firing Atlantis's braking rockets at 7:45 A.M. Eastern Daylight Time on July 7, the pilots brought Atlantis home to runway 15 at Kennedy's Shuttle Landing Facility. About an hour after landing, Mir-18 cosmonauts Dezhurov and Strekalov and U.S. astronaut Thagard were brought out of the Shuttle into the Crew Transport Vehicle alongside Atlantis for their initial postflight medical testing. After having spent 115 days in space, the crew members had difficulty readjusting to an environment containing gravity. The three were then flown in an Air Force C-9 Medevac plane to Ellington Field in Houston, Texas, for several weeks of medical tests and physical therapy to regain their strength, muscle tone, and stamina.

So what does all of this imply? The significance seems to rest on the international context of the Atlantis-Mir docking mission and what it signaled for the future of spaceflight. Humans eventually may look back on this flight as the beginning of a cooperative effort that created a presence for Earthlings beyond the planet. On the other hand, maybe it will prove to be only a minor respite in the competition between nations for economic and political supremacy. Perhaps most important is that humans today have the opportunity to create an international space station that could enable the future to move off of this planet.[23]

NASA undertook nine Shuttle docking missions to Mir between 1995 and 1998. Some were stunning in their scientific return, others proceeded from disaster to disaster. Often it appeared that the crew might be in harm's way.

CRISIS ON MIR

Most of the time that American astronauts spent aboard Mir was routine, even boring. Shannon Lucid's experience, seemingly idyllic, involved day-to-day tasks pertaining to scientific experiments. She had considerable off time to read books, a pastime she had missed for several years. "It really sort of brought home the power that authors have . . . here I was . . . reading *David Copperfield* and *Bleak House.* I thought, 'Wow, here was this guy [author Charles Dickens] that lived in a totally different era than we're living, and it had never ever crossed his mind that his book would be . . . read . . . by an American, on a Russian space station.'" She added, "I mean, that would have just absolutely blown his mind, that the words that he penned way back there in England, I was reading. I just thought about that a lot, about the power that authors have and his ideas and his story was transcending the centuries, transcending culture."[24] But Lucid's quiet hours on Mir pointed up the nature of several life-threatening incidents aboard the space station. When she left, most of her books stayed behind for others to enjoy. They were stored in the Spektr module in June 1997, when a Progress resupply vehicle rammed it and rendered it inaccessible, nearly destroying Mir altogether.

The succession of 1997 failures probably had more to do with age than anything else. Designed to last five years, the station had by then been in space for eleven and was literally falling apart. Warning alarms went off regularly. Hoses split, releasing antifreeze that the astronauts had to breathe in. Devices broke down. There were numerous power failures. Garbage and broken equipment built up because there was not enough room in the spacecraft to get rid of it.[25]

The first serious crisis took place on February 24, 1997, when astronaut Jerry Linenger and his fellow crew members fought a fire ignited by a malfunctioning oxygen generator in Kvant 1.[26] Although the fire burned for only about ninety seconds, the crew was exposed to heavy smoke for five to seven minutes and donned masks in response. Linenger had been in the Spektr module working on his computer when he heard Mir's master alarm go off. He shut down his computer—in case the power should go off—put on some protective gear, and rushed as best he could in his weightless condition to the scene of the accident. They all realized that the fire was serious; it could jeopardize the station and their lives, for it blocked access to one of the Soyuz spacecraft needed for return to Earth. Until it was extinguished, only three of the six men on the station could escape. They put out the fire with foam from three fire extinguishers, each containing two liters of a water-based liquid.[27] The blaze was not small. Burning in all directions in the microgravity of the space station, oxygen from the generator fueled hydra-like flames up to three feet long. Periodically, said Linenger, bits of molten metal from the oxygen generator were sent "flying across and splattering the other bulkhead."[28]

The station quickly filled with smoke, and that proved an even greater hazard. Linenger recalled, "The smoke was the most surprising thing to me. I did not expect smoke to spread so quickly." It moved at "a magnitude about ten times faster than I would ever expect a fire to spread on a space station. The smoke was immediate. It was dense. . . . I could see the five fingers on my hand, I could see a shadowy figure of the person in front of me who I was try-

Cosmonaut Yuri Gidzenko ponders the clutter of the Mir station in 1995. (NASA photo, no. STS074-322-014)

ing to monitor to make sure he was doing okay, but I really could not make him out."[29] Once the flames had been contained, they started purging the atmosphere of the smoke. Linenger, a physician, examined the others. They wore masks and goggles until an analysis of the atmosphere ensured they would experience no serious health risk.

Russian officials soon ascertained that the problem had begun when a crack in the oxygen generator's shell allowed the contents of the cartridge to leak into the hardware in which it was located. The damage resulted from excessive heat rather than open flame. The crew also reported that the heat melted the outer insulation layers on various cables. Mir was a hardy little vessel, however, and its major systems continued to operate effectively. "It is unfortunate that this incident occurred, but we are thankful that there were no injuries," Frank Culbertson, director of the Phase One Shuttle-Mir program, said at the time. "Russian management and operations specialists have been very informative as to what happened, and we are working closely with them on evaluating the health of the crew and how best to respond to the damage," he added. "The crew did a great job handling the fire, and the ground support has been excellent on both sides."[30]

Immediately after the fire, the crew spent the entire day mopping up residue from extinguishers and smoke, wiping down bulkheads, cleaning the cabin, and putting damaged equipment into bags. They did not conduct any research that day, but a day or so later, Linenger returned to his science mission and carried on where he had left off. The Shuttle-Mir program manager offered several lessons from this experience:

> The number one lesson is that the critical people in a situation like this are the crew on-board. In this case, I think all of the crewmembers responded in an outstanding fashion. They obviously were well trained, they responded quickly and confidently and did all the right things very fast in order to control the situation on the Mir and the cleanup afterwards. I think we are learning some facts that we need to get into our hopper on what do we do if we have a situation like this on a station, and we're not abandoning it.
>
> On the shuttle if we have a major problem like this we bring the shuttle home and repair it on the ground, but with a station you may or may not want to send the crew home, and if it's not something that you would abandon the station for but want to continue operations, you need to know how you're going to clean up after that and continue to do the operations down the road. We're looking at that, among other things, including what kind of equipment they used for breathing during the smoky period and how they put the fire out.[31]

Perhaps the most troubling aspect of this incident was that the Russians were slow to inform NASA of what had taken place on Mir. When NASA officials realized that an accident had taken place, the Russians minimized its seriousness, claiming that the fire had burned itself out and that the station could easily be cleaned up. The damage was also downplayed, and the astronauts were blamed. NASA went along with this attitude for the sake of international relations, but mistrust of the Russians, already growing for some time, exploded after the incident. Looking back, a senior space station official remarked that from this point forward, he accepted nothing the Russians told him without independent verification.[32]

The fire foreshadowed a series of problems aboard Mir during the spring and summer of 1997. Oxygen generators broke down, the automatic docking system malfunctioned, various types of equipment (both great and small) interrupted the normally monotonous activities, the station's orientation system broke down, the power system failed when the solar arrays lost their position toward the Sun, and leaks in the Kvant-2 cooling system forced numerous repairs and seemingly endless fussing to keep it running. It appeared that the Mir crew, including Linenger, spent the majority of their days repairing the space station. They gingerly positioned Mir in relation to the Sun so that they could control temperature on various parts of the station. The environment was uncomfortable and the crew complained about it, but Linenger commented in retrospect, "We're out here in the frontier and I guess I expected the unexpected. We've been getting some of that. . . . So, in spite of some of the difficulties, we've been having a very successful mission. And, some of the system problems—I can't say that I expected them. But . . . I was trained to work on those systems and assist the crew where I could."[33]

Linenger believed that Russian mission control may have kept him and his Russian comrades in constant danger, because the crew was not informed about the status of their station. As for his fellow crew members, he expressed nothing but praise for their strength and perseverance throughout the mission. Even with communication difficulties, a cloud of doubt surrounding the station's systems, difficulties with mission control, and fires and toxic fumes, the crew worked well together. Jerry Linenger returned to Earth aboard the Space Shuttle Atlantis, which docked with Mir on May 16, 1997.[34]

The problems Linenger faced on Mir were nothing compared with what astronaut C. Michael Foale and two Russian cosmonauts experienced on June 25, 1997. On that date, a Progress resupply vehicle rammed the Spektr module and caused rapid depressurization of the module along with the loss of several critical electronic systems on Mir. Without question this accident was the most serious to take place in space since the near tragedy of Apollo 13 in April 1970, when an oxygen tank exploded en route to the Moon and the crew returned home safely only after heroic efforts on the part of the entire NASA flight team.[35]

The 1997 accident occurred during an attempted manual docking of the Progress. Cosmonaut Vasily Tsibliev, at the remote controls operating Progress from the station, fired its rockets to propel the craft toward Mir. Much as in an arcade game, Tsibliev "flew" the Progress from onboard Mir while he watched the episode on a video screen. Tsibliev had trouble making out the station from the Progress's video readout, because it got lost in the background of Earth's cloud formations below. Mike Foale recalled, "What Vasily was seeing on his screen was an image that didn't change in size very fast. That's the nature of using a TV screen to judge your speed and your distance. He couldn't determine accurately from the image that the speed was too high."[36]

When Aleksandr Lazutkin realized what was happening, he yelled to Foale, "Michael, get in the escape ship!" Lazutkin recalled, "I watched this black body covered in spots sliding past

Astronaut Shannon Lucid exercises on a treadmill that was placed in the Mir base block module during her extended stay in 1996. (NASA photo, no. NM21-399-001)

below me. I looked closer, and at that point there was a great thump and the whole station shook." The Progress collided first with a solar array on the Spektr module, crumpling it like a fragile dragonfly wing before hitting Spektr itself. The crash destroyed solar panels, buckled a radiator, and breached the integrity of Spektr's hull. The module decompressed, but the crew was uninjured. The accident crippled the station and led to a series of crises in space.[37]

All of Mir immediately began losing air pressure, approaching the 540 millibars necessary for safety. The crew sealed the node between Spektr and the rest of the space station, but not without difficulty because of all the cables and conduits running through the hatches. These had to be removed first and the hatch sealed. At the same time, Mir went into a spin when the station's gyro shut down and attitude control failed. The crew worked to stop the spin and face the solar arrays back toward the Sun. Meantime, the station looked more like a sinking hulk at sea than an operational vessel, as it listed and spun without control. With inputs from the crew, ground controllers in Russia fired Mir's engines to stabilize the station's attitude.[38]

For the next two days, the crew operated without power and under enormously taxing conditions. They then spent the rest of the summer recovering as best they could. The Russians launched another Progress resupply ship to take up critically needed supplies and repair materials. That cargo craft, operated through automatically controlled systems, docked without incident on July 7, 1997. Along with food, fuel, and clothing, it carried special supplies for an internal spacewalk to recover the use of the solar arrays on the station's Spektr module. Lines bringing power from them into the main compartment had been disconnected when the module was sealed following the collision. Now the crew successfully installed a modified hatch, through which power lines could be routed while keeping the module sealed. They also continued to work to restore systems to Mir.[39]

On September 6, 1997, Foale and Commander Anatoly Solovyev conducted a six-hour EVA to inspect damage to the station's Spektr module. Foale's scientific activities aboard Mir had been interrupted by the Progress accident, and since most of his experiments were aboard Spektr, he was unable to complete all of his tasks. Foale returned to Earth on October 6, 1997, with the crew of STS-86 aboard Space Shuttle Atlantis.[40]

The near loss of Mir and its crew triggered criticism of U.S.-Russian cooperation in the Shuttle-Mir program. That criticism emphasized the safety of the astronauts and the aging character of Mir, noting that it was in such bad shape that no meaningful scientific results could be achieved aboard the station. Most American space advocates asked Russia to deorbit Mir and thereby bring it to "an honorable end." Others thought the Russians had engaged in an irresponsible action by flying Progress manually, believing that the equipment was suspect in the first place and the Russian Space Agency had needlessly placed the crew in jeopardy. Critics in the U.S. Congress demanded that NASA terminate the cooperative program immediately, but administrator Daniel S. Goldin insisted that Mir was overall quite safe, and that the Shuttle-Mir program should continue. The Russian Space Agency managed to keep the station operational, but it was never the same thereafter.[41]

Two more NASA astronauts took up residence on Mir in 1997 and 1998. In June 1998 the last one, Andrew Thomas, returned to Earth aboard Discovery, STS-91. The operational phase of the Shuttle-Mir cooperative mission ended on August 25, 1998, when the Mir-25 cosmonauts brought back the final science results and turned them over to NASA officials. It was the end of a highly controversial program. In the end, no fictional thriller could compare with the true story of the snowballing of accidents, mechanical failures, and near catastrophes that occurred on Mir in 1997 while Jerry Linenger and Michael Foale were aboard.

In retrospect, how should the Shuttle-Mir program be characterized? What did NASA learn about long-duration human spaceflight, international cooperation in space, and the long-term effort to create a base camp to the stars? The program had originated as a "dress rehearsal" for a partnership to build an International Space Station, but failures of the dilapidated, eleven-year-old Mir made it seem a poor decision. The Russian-American crew endured six months of nonstop crises, including fire, power blackouts, chemical leaks, docking failures, and constant breakdowns. Perhaps the most important lesson learned in the Shuttle-Mir program was that technology was not the main challenge to space exploration. Rather, it was the human factor, the ability of people to interact effectively with each other.

Lessons to be gleaned from these events pertain to the awful consequences of not considering the human element in long-duration space operations and the fact that crews of depressed, isolated, overworked, and scared people make mistakes. We can learn from the collision of culture and politics in spaceflight and from the consequences that resulted when technical decisions were made for political reasons. It quickly became clear during the Shuttle-Mir program that both NASA and the Russian Space Agency were ill prepared to cooperate effectively in space. The Americans had little knowledge of Russian capabilities for spaceflight, even less of the spacecraft and operational processes used. The Russians had similar problems and seriously misconceived ideas about the partnership. Thus both countries immediately encountered problems working together.

Historically, Shuttle-Mir shares similarities with all of NASA's earlier major human spaceflight programs. As in Project Mercury, NASA found that the individual astronauts received the majority of public attention. For many years astronauts had been faceless overachievers without distinct personalities, perhaps because there were so many of them, with Space Shuttle crews of up to seven members. Few Americans could recite their names. But the Mir astronauts, each of whom emerged as distinct public personalities, supplanted these doughfaced astronauts. Americans found themselves identifying with the experience of aloneness felt by Norm Thagard and Shannon Lucid and the dangers on Mir experienced by Jerry Linenger and Michael Foale. The emergence of these astronauts from their faceless coterie conjured up the popularity of John Glenn and the other members of the seven Mercury astronauts chosen in 1959.

Like Project Gemini, Shuttle-Mir served as a critical stepping stone to a higher goal. Gemini had been established to learn skills necessary to land on the Moon, especially lengthy missions of up to two weeks, rendezvous and docking of spacecraft in flight, and extravehicular activity. Shuttle-Mir was devised to prepare for the much more ambitious International Space Station program. Carrying the Gemini analogy further, Shuttle-Mir astronauts practiced the same skills, extending American long-duration spaceflight records far beyond anything experienced before, engaging in new rendezvous and spacewalking techniques required for the International Space Station. Like the Gemini crew of Neil Armstrong and David Scott, the Shuttle-Mir astronauts also had to improvise to rescue themselves and their craft from serious, life-threatening malfunctions.

At a fundamental level the Shuttle-Mir program served the larger purpose of enhancing international relations between the United States and Russia, which had for half a century been strained in an intense Cold War. "It was a very emotional experience," said Tommy Capps, an engineer working in the astronaut-training program. "Really, it was a sobering experience for me—because I really got to thinking about how far we had come." Earlier in the program, "We would have had a lot of sidestepping, and sashaying back and forth."

But by STS-91, as NASA's Shuttle-Mir program manager Frank Culbertson said,

We were there. We were clicking. And we were a program. We had no idea how much we would learn. I think the most valuable lesson is that you're going to have things happen that are going to require problem solving continuously. The best-laid plans, the best-designed systems—you're still going to have difficulties. Long-duration spaceflight is hard, and the most valuable lesson we learned from that is to expect it to be difficult. Plan for that. Train for that. And then be prepared to handle the unexpected.[42]

Shuttle commander Charles Precourt added:

What I hope the American public can glean from the Shuttle-Mir Program is that all hardware breaks down. We have to learn to take our hardware to space and not bring it home in a hurry—like an airliner that might be flying home that has a problem. If we don't learn to do it out there, we won't ever be able to stay there very long. Because hardware does fail. Our ultimate goal is to be able to go to the Moon and Mars, and put bases there for scientific research and for exploration purposes, and stay there and survive. The fact that the Mir went through ups and downs and we were able to live through that . . . is a great testament to what we were able to do together, and it should make people think twice when they try to criticize the Russians and their system for what it is or is not capable of doing.[43]

"Mir operations and life were hard," Frank Culbertson said. "A lot of people have said it's the hardest thing they've ever done. And my prediction is that the [International Space Station] is going to be even harder, for a lot of reasons that people don't understand yet. . . . So, this next 10 years is going to be really, really fascinating."[44]

Most important, this program demonstrated that large-scale space exploration efforts will no longer be carried out as a competition between nations. NASA officials pointed out that it yielded some enormously significant results:

- Developed flexibility by operating in space with several launch vehicles and a space station
- Conducted long-term operations with multiple control centers and U.S. and Russian teams at each other's facilities
- Conducted spacewalks outside both Russian and U.S. vehicles with astronauts and cosmonauts testing each other's space suits, a preparation for joint walks to assemble the International Space Station
- Trained astronauts, cosmonauts, and other team members in each other's language, methods, and tools to facilitate operations in orbit and make mission training more efficient
- Created a joint U.S.-Russian process for analysis, safety assessment, and certification of flight readiness
- Led to refinements in software, hardware, and procedures that will be used in operations onboard the International Space Station
- Established that noncritical systems may fail and be replaced through routine maintenance, without compromising safety or mission success
- Showed that multiple oxygen-generation systems are essential for safe, uninterrupted operations
- Mated U.S. and Russian hardware in orbit and verified complex robotics operations during delivery and assembly of the Russian-built docking module
- Collected data on the effects of long-duration exposure of hardware to the space environment
- Learned how to conduct long-term research and maintenance on a space station through flexible scheduling of crew time on orbit
- Developed a process for mission planning and psychological support for astronauts on orbit during extended periods[45]

THE DEORBITING OF SPACE STATION MIR

Located a mere 250 miles above Earth, Mir required repeated reboostings to maintain its orbit. These, of course, ensured that eventually it would be abandoned, reenter the atmosphere, and burn up as it plunged to Earth. Space station Mir, then the heaviest object orbiting this planet other than the Moon itself, returned to Earth early in the morning of March 23, 2001, after 86,331 total orbits. Five of Mir's modules were still pressurized with air at the time of the deorbit. When they exploded, sky watchers (mainly sea birds) witnessed incandescent fragments streaking across the horizon. Some of the larger pieces blazed into the sea about 1,800 miles east of New Zealand, and observers in Fiji reported spectacular gold and white streaming lights. Fragments from the massive complex splashed down in the South Pacific Ocean just as ground controllers had planned. No one was hurt. On the contrary, onlookers described the experience of a lifetime. Of the station's 135 tons, only about 20 reached the surface, most of them in small pieces.

During its fifteen-year existence, Mir had set endurance and space-adventure records that will stand for some time. The nations involved in the building of the International Space Station interpreted the deorbit as a step forward. Mir's lifetime was significant, but it was an interlude leading toward using the ISS to undertake long-term research into the questions requiring resolution before humans could be sent on a long-term basis to the Moon and then on to Mars.[46]

Concerns circled the globe about Mir crashing into populated areas. The space station's path crossed over nearly every continent on Earth. Its orbits tracked over everything between 51° North and South latitude, roughly within the limits of the Aleutian Islands to the north and the southern Andes Mountains to the south. Pieces of previous large spacecraft had landed in Canada, Australia, and southern South America, fortunately without damages or casualties. The U.S. government provided Russia with tracking and trajectory data, atmospheric conditions, and even solar activity that might cause Earth's atmosphere to expand farther into space. Although there was considerable certainty that debris could be limited to falling in the ocean, Russian space official Yuri Semenov, president of RSC Energia (which to the Russian Space Agency is as the Kennedy Space Center is to NASA), was quoted as saying, "We don't have a 100 percent safety guarantee."[47]

But Mir returned without incident. An amazing saga and highly successful program had come to a watery end. Anatoly Solovyev had lived a total of 651 days on Mir and served as Mir-24 commander for several American astronauts. Solovyev was quoted as saying, "I am especially sad these days. An entire era of our Soviet space program is ending, into which we invested not only our money but, what is more important, our intellectual potential."[48]

The International Space Station was well under way by the time of the Mir deorbit, but the memories of Mir refused to fade. Indeed, some embraced its demise as the stuff of legend. In a 1998 interview, Vladimir Semyachkin reflected on Mir. He had developed the motion control systems and navigation systems for all vehicles and stations that were produced and launched into space by RSC Energia. Semyachkin, as much as anyone, had wrestled with Mir's problems. He said: "It's a shame. . . . Our child, who we gave birth to so many years ago . . . we're going to have to put it to sleep. But, on the other hand, we understand that sometimes there's nothing to be done. . . . One cannot sit, as it were, on two chairs at the same time. Nevertheless, despite this sorrow with . . . regard to Mir, we nonetheless do look forward to the future with a great deal of hope."[49] At a fundamental level, the demise of Mir made way for the International Space Station, the first elements of which reached orbit in the fall of 1998.

CHAPTER 6

Building the International Space Station

Beginning in 1993, the National Aeronautics and Space Administration transformed the Space Station Freedom program into the International Space Station. NASA leaders were thereafter remarkably successful in maintaining the political coalition supporting the effort, and in late 1998 the station's first elements were launched into orbit. Assembly continued throughout 2002, and at this point NASA expects to complete it by 2006. The first crew went aboard in the fall of 2000; a total of five crews had served aboard this station by September 2002. Although there have been enormous difficulties—cost overruns, questions about the quality of science to be undertaken, the role of civilians who want to fly—one may appropriately conclude that the ISS effort has thus far been successful. This is true for three major reasons.

First, the fact that this station is being built by a large international consortium is extraordinary, given the technical, financial, and political obstacles involved. The U.S. House of Representatives came within a single vote of canceling the entire effort in 1993. Space organizations from a multitude of nations have struggled to overcome cultural differences on this enormously complex high-technology undertaking.

Second, the International Space Station provides the most sophisticated model ever offered for tax-financed human activities in space. One hundred years hence, people may well look back on its construction as the first truly international endeavor among peaceful nations. And third, the station may be able to revitalize the spacefaring dream. Once functioning in space, it may energize the development of private orbital laboratories. Such labs could travel in paths near the International Space Station. The high-tech tenants of this orbital "research park" could well take advantage of the unique features of microgravity and achieve remarkable results. The ISS would permit research not possible on Earth in such areas as materials science, fluid physics, combustion science, and biotechnology.[1]

From Space Station Freedom to the International Space Station

In the summer of 1993, the Space Station Freedom program stood at a crossroads. A U.S.-led international effort to build and operate a permanently occupied Earth-orbiting research facility, Freedom, had been designed to play several roles. It would become an orbital scientific laboratory for microgravity, Earth observation, and other scientific experiments. It would also serve as a facility to study and develop skills for long-term human survival in space, a rather quietly whispered role for Freedom. Finally, it would be a model for international cooperation.[2]

The program began officially in January 1984, when President Ronald Reagan announced the U.S. intention to build a space station. The original plan called for Freedom to be built by the early 1990s, but funding concerns prompted restructuring without really resolving its inherent budgetary and technical problems. These program redesigns and funding reductions sent the Freedom program into a tailspin from which it never recovered.[3] The Congressional Budget Office decided in the spring of 1993 that Freedom should lead a list of options for trimming domestic discretionary spending. Its cancellation, budgeteers believed, would save the government $10.4 billion over five years.[4]

The United States was responsible for the vast majority of the station budget. It spent about $10 billion on pre-Alpha station work and was to spend an additional $28 billion on design, con-

This rendition of the International Space Station, prepared in 1993, shows the U.S. and Russian human-tended elements that would be on orbit during Phase 2 of the cooperative program. (NASA photo, no. 93-HC-431)

struction, launch, and assembly to complete the station. The Japanese anticipated spending $3 billion on the Japanese Experimental Module. ESA considered a $3 billion station-related program, and Canada was to spend about $1 billion.

In such a climate, Space Station Freedom seemed doomed. In mid-March 1993, incoming president William J. Clinton directed NASA to downsize Freedom for a fourth time. Responding to this call, NASA came forward with three options in June 1993. These were a modular concept that would use existing flight-proven hardware; a derivative of the current Space Station Freedom design; and a space station that could be placed into orbit with a single launch of a

Shuttle-derived vehicle. Each of these designs achieved several major objectives that NASA administrator Daniel S. Goldin had emphasized in his charge to the redesign team:

- Provide a cost-effective solution to basic and applied research challenges whose merit is clearly indicated by scientific peer review, significant industrial cost sharing or other widely accepted method
- Provide the capability for significant long-duration space research in materials and life sciences during this decade
- Bring both near-term and long-term annual funding requirements within the constraints of the budget
- Continue to accommodate and encourage international participation, and reduce technical and programmatic risk to acceptable levels

In addition, the team worked in an exceptionally constrained funding environment. For the five-year period from fiscal years 1994 to 1998, NASA was to assume a low option of $5 billion, a mid-range option of $7 billion, and a high option of $9 billion. Within that budget it had to fit all space station development, operations, utilization, Shuttle integration, facilities, research and development, and an appropriate level of reserves.[5]

On June 17, 1993, the president announced that although the United States would continue a space station program, honoring its commitment to thirteen international partners, it would be a much-pared effort. He chose the Alpha design proposed in the redesign effort, a medium-sized modular space station based on work already done. Using a combination of Freedom hardware and flight-qualified space systems from other sources, the Alpha design represented the middle path to developing the station. Alpha had four distinct phases:

1. Photovoltaic (PV) Power Station on-orbit for increased power to a docked orbiter/ pacelab
2. Human Tended Capability (adding a U.S. laboratory)
3. International Human Tended (adding an additional PV array and international modules)
4. Permanent Human Capability (adding a third PV array, the U.S. habitat module, and two Russian Soyuz capsules)

This gradual approach would, according to the NASA statement at the time:

- Make maximum use of Space Station Freedom systems and components, which have completed a rigorous critical design review, where it is both cost effective and schedule

enhancing, thus benefiting from the nation's investment to date in the Freedom program
- Incorporate changes that would reduce complexity and increase the probability for meeting cost and schedule
- Achieve substantial savings through a streamlined management structure that provides clear lines of authority, reduces overlap and gives accountability and authority to the lowest level to get the job done; and
- Benefit from a new operations approach that would significantly reduce operational costs[6]

Goldin announced that Russian hardware alternatives, where they could benefit the re-designed program, also had been incorporated into the plans. The Soyuz crew return vehicle, for instance, could be used until a new system had been developed, and Russian launch and other systems might suffice to reach the new station. At the same time that President Clinton announced the Alpha concept, he also planned to "seek to enhance and expand the opportunities for international participation in the space station project, so that the space station can serve as a model of nations coming together in peaceful cooperation."[7]

The Alpha design was, most importantly, a downsized representation of Space Station Freedom with the capability for a crew of only four, totally reliant on the Space Shuttle for transportation and supply. The original redesign options had projected Permanent Human Capability (PHC) in 2001 or 2002. But these scenarios also required peak annual funding levels of at least $2.8 billion. In June 1993 Congress and the Clinton administration directed NASA to keep the program within an annual expenditure level of $2.1 billion. NASA reassessed the assembly plans given this constraint and revised the schedule for achieving PHC, deferring it to September 2003. Independent assessments set the cost of Space Station Alpha at $19.4 billion for the ten-year period.[8]

NASA leaders embraced this plan and quickly wrote off Freedom in favor of a nearly clean slate. The slate became possible only because of the collapse of the Soviet Union in 1991 and the desire to rope the former Communist state into closer ties to the West. At a Vancouver summit in April 1993, the Clinton administration invited Russia to participate in a renewed space station program, and Russian president Boris Yeltsin agreed. The new International Space Station project, based on a downsized Alpha design, called for thirty-four construction-related Space Shuttle flights. NASA developed a three-phase approach to the program:

Phase 2, 1994 to 1997: Joint Space Shuttle/Mir program
Phase 2, 1998 to 2000: Building of a station "core" using U.S. node, lab module, central truss and control moment gyros, and interface to Shuttle; Russian propulsion, initial power

In 1995 the first elements of the International Space Station began to be constructed. Here the Unity node of the ISS is nearing completion. Finished in spring 1997, Unity was shipped from the Boeing manufacturing facility at Marshall Space Flight Center, Alabama, to the Kennedy Space Center, Florida. There it began launch preparations in June 1997. Although the 22-foot-long by 14-foot-diameter node was essentially a passive station component, 216 lines for fluids and gases, 121 electrical cables, and 6 miles of wire were installed to provide connections to other modules. Unity was launched aboard the Space Shuttle Endeavor in December 1998 and was docked with the already orbiting Zarya module, launched aboard a Russian Proton rocket. (NASA photo, no. 95-HC-142)

system, interface to Russian vehicles, and assured crew-return vehicle; Canadian remote manipulator arm

Phase 3, 2000 to 2004: Station completion. Addition of U.S. modules, power system, and attitude control; and Russian, Japanese, and ESA research modules and equipment[9]

The Challenge of Russian Cooperation

American collaboration in the International Space Station caused resentment among the U.S. partners. One space policy analyst believed the Europeans and Japanese saw the U.S. position as "arrogant and, particularly in Europe, insufficiently insensitive to a partner's ability to contribute significantly to the station program." The foreign partners were further dismayed, he felt, by official NASA statements that the space station was critical to U.S. leadership and that international collaboration would "engage resources that otherwise might be used in support of programs competitive to the United States." This U.S.-dominating philosophy conflicted with fundamental European and Japanese desires to achieve their own areas of autonomy in space programs and equal technical cooperation with the United States.[10] This made it more difficult to forge commitments among partners and to reach detailed agreements on management and utilization issues.[11] A 1989 NASA internal design review excluded the space station's foreign partners, thus causing further tension in the cooperative relationship. After 1990 NASA made a greater effort to include partners in station redesign activities, but problems still abounded. An Office of Technology Assessment concluded, "The space station experience appears to have convinced the partners that they should not enter into such an asymmetrical arrangement [with the United States] again."[12]

With the addition of Russia as a station partner in 1993, however, the U.S. position on collaboration changed fundamentally. Under the new International Space Station program, the United States would rely on Russia for several critical elements, including guidance, navigation, and control in Phase 2; habitation until the U.S. habitation module was launched; crew return (lifeboat) through 2002; and reboost and fuel resupply. The Russian collaboration policy evoked high levels of controversy in the United States and among the space station's other foreign partners. Domestic objections to dependence on Russian technology were based on concerns about supply. Russia's political and economic stability, questions about its technical reliability, the potential for loss of U.S. jobs, and traditional pressures to maintain U.S. control over critical mission elements all affected the program.[13] Moreover, partners from other countries expressed resentment over not having been consulted about Russia's sudden entry into the program.[14]

To proceed with the Russian involvement in the new ISS, on January 13, 1995, NASA signed a $5.63 billion contract with Boeing to manage the building of the core station, including two nodes, an airlock, and laboratory and habitation modules, as well as their integration. Boeing immediately forged a strategic alliance with the Russian Khrunichev design organization and announced a subcontract to build the Functional Cargo Block, which in the Cyrillic alphabet yielded the acronym FGB. In essence, the Russian Space Agency became a contractor for the FGB module. The cost advertised at the time was $190 million, but that soon grew to $210 million. It was the first of several such increases.[15]

Because of its unique circumstances, Russian cooperation quickly became a different experience from that of the other international partners. Whereas Europe and Japan contributed add-on pressurized research modules and other plug-in components, the Russians were to provide several critical station modules without which the system could not function. These included FGB (for guidance, navigation, and control), reboost and refueling, a service module, a power mast, and Soyuz capsules for emergency return.[16] All of this placed Russia firmly in the critical path of the space station program. If the Khrunichev design bureau failed to perform, it could derail the entire effort. This made some uncomfortable, despite the perceived need of the Clinton administration to emphasize strategic relations with the United States' former Cold War rival.

Beginning in the 1980s, for several reasons the United States had to bring other nations increasingly into the core of collaborative space projects. U.S. preeminence in space technology was coming to an end, as ESA developed and made operational its superb Ariane launcher. Advancing space capabilities made it increasingly possible for Europe to "go it alone" unless the United States made its international collaboration more integral to space exploration.[17] Additionally, U.S. commitment to sustained preeminence in space activities waned, and significantly less public monies went into NASA missions.[18] Finally, these realities increasingly led NASA to accede to the demands of international collaborators to develop their own critical systems and technologies. This overturned the policy of not allowing partners into the "critical path," something that had been flirted with but not accepted in the Space Shuttle development project.[19] It was in large measure a pragmatic decision on the part of American officials. Because of growing size and complexity, according to space policy analyst Kenneth Pedersen, recent projects had produced "numerous critical paths whose upkeep costs alone will defeat U.S. efforts to control and supply them." He added, "It seems unrealistic today to believe that other nations possessing advanced technical capabilities and harboring their own economic competitiveness objectives will be amenable to funding and developing only ancillary systems."[20]

At the same time, several observers voiced concerns about the use of NASA funds as a foreign aid program for Russia. With the decision that NASA would contract with Russian space organizations for critical components for the International Space Station, the Clinton administration allowed economic and foreign policy interests to manipulate the effort. In an unprecedented

Russian technicians work in September 1997 on the almost completed forward portion of the U.S.-funded Functional Cargo Block (Russian acronym FGB), Zarya, built by the Russian Space Agency. Russia built the module under contract with the United States, and the FGB began prelaunch testing shortly after this photo was taken. The forward docking area shown on the left will be docked with the U.S.-built Unity during the first Space Shuttle assembly mission aboard Endeavour, December 1998. (NASA photo, no. S97-12609)

decision, the United States agreed to spend nearly $650 million in direct payments to Russia for procurement of equipment for the station. In the words of Keith Cowing, director of watchdog organization NASA Watch, this represented "the abduction of [the] space station program by the State Department for use as a conduit for foreign aid to Russia." Cowing maintained:

> Instead of removing Russia from the critical path, the White House has stood by and allowed Russian non-performance to delay progress and drive up costs. Then, to compound this situation, the White House has the gall to tell NASA that these problems all have to be solved within NASA's existing—and shrinking—budget. The net result is that all of NASA's other programs are, or will be, delayed, shrunken, or abandoned.
>
> The net result is that the space station program has technical decisions being made for political reasons and political decisions being made for technical reasons.[21]

Although Cowing's position was more extreme than most, his comments resonated with those of many who were knowledgeable about the International Space Station program and the 1993 decision to bring in the Russians as the most significant partners in the effort.

Cowing's views were confirmed in less strident language by Library of Congress space policy analyst Marcia Smith. She said in testimony before Congress in 2001:

> From the beginning, challenges arose with Russia's participation. Many promises were made by high ranking Russian government officials that sufficient funding would be provided to fulfill Russian commitments to ISS. Most were not kept. NASA initiated contingency plans to cope with the possibility that Russia might not build or launch elements that were in the critical path, including the Zvezda Service Module which provides crew quarters and guidance, navigation and control (GN&C) functions. Russia's ability to provide sufficient Soyuz "lifeboat" spacecraft and Progress "reboost" spacecraft also was questioned. Funding for Russia's space program was under severe stress, and construction of Zvezda, for which Russia was expected to pay, was significantly delayed.[22]

These concerns required redress, and NASA took action to hedge against Russia's nonperformance.

In April 1997 NASA reallocated $200 million in fiscal year 1997 funds from the Human Space Flight Program to a new budget line item called Russian Program Assurance. In its original version, NASA suggested that the fund could also be used to develop contingency alternatives to buy down the cost and schedule risks that might result from other Russian shortfalls as well as those in this specific program, but Congress approved only the leaner budget. With it NASA was able to contract with the Naval Research Laboratory to build what they called an interim control module, viewed as necessary if the Russians failed to make good on their commitment.[23]

THE STRANGE SAGA OF ZARYA: THE FUNCTIONAL CARGO BLOCK (FGB)

Throughout the construction of ISS, the Russian efforts always seemed more over budget and behind schedule than those of other partners. The first instance of conflict came with the U.S.-funded and Russian-built Functional Cargo Block control module (FGB), named Zarya ("sunrise") when the Russians finally launched it on November 20, 1998. Zarya was a critical component that had to be sent into space before anything else. The 42,600 pound, 41.2 feet long, and 13.5 feet wide pressurized module, often referred to as "tugboat" because of its control capability, was to provide orientation control, communications, and electrical propulsion for the station until the launch of the third component, Russian-provided crew quarters. Later during assembly, Zarya became a station passageway, docking port, and fuel tank. Zarya was built by Russia under contract to the United States and is owned by the United States. The FGB sported two solar arrays 35 feet long and 11 feet wide. Its sixteen fuel tanks could hold more than six tons of propellant, providing attitude control for the station with twenty-four large and twelve small steering jets. Two large engines were also available for reboosting the spacecraft and making major orbital changes.[24] To say the least, the FGB proved a critical, and divisive, component of the ISS project.

In December 1995 the Russian Space Agency informed NASA that it would be unable to complete the FGB on time and on the original budget of $190 million. With the FGB in the critical path of the project, the Russians offered an alternative that made all the other partners irate. Instead of building a totally new FGB, they offered to use Mir as the core of ISS. Such an offer would get Russia off the hook for a very complex FGB program. After all Mir was already built, in orbit, and operational. But it was also a decade old and having seemingly endless systemic problems. Mir had previously been offered as the ISS core in 1993, and the proposal had been roundly rejected. Accepting it now would be no more appealing. NASA sent repeated entourages of negotiators to Moscow to meet with their RSA counterparts. They achieved little success in budging the Russians from their position that the FGB would be produced behind schedule and over budget, if at all. Representatives James Sensenbrenner (Republican, Wisconsin) and Jerry Lewis (Republican, California) of the House Science Committee also went to Moscow and met not with RSA officials but with the nation's chief officers. The result of their visit, they later learned, was that the FGB would be built on time, although additional assistance would be required. If the Shuttle could take up some of the launch requirements needed to meet the Shuttle-Mir and ISS assembly program, RSA could concentrate on the FGB. NASA leaders collectively gritted their teeth as they signed this revised agreement, feeling as if they had been forced to accept an unreasonable deal to preserve the greater ISS partnership.[25]

No sooner had the ink dried on this 1996 agreement than the Russians announced that the schedule could still not be met. The launch of the FGB had to be deferred by seven months, from November 1997 to June 1998. In September 1997, after coordination with the international partners on future ISS assembly flights, NASA developed a new manifest that reflected a slip of more than a year in completion of ISS assembly. These events and others raised questions regarding the total cost and schedule for ISS development and operations. The FGB module finally reached orbit in November 1998.[26] A report on ISS cost and schedule noted in April 1998:

> The anticipated one billion dollar cost savings to the U.S. to be accrued from Russian provision of the Functional Cargo Block (FGB in its Russian language acronym) and an Assured Crew Return Vehicle capability, was a faulty assumption as far back as 1994. The continuing economic situation in Russia has also negated most of the $1.5 billion in schedule savings to be achieved through their involvement. Russian schedule slippage, due largely to failure of the Russian government to deliver promised funding, translates directly to the most recent Service Module schedule slips. With continuing funding shortfalls carrying into 1998, the absence of any hard indicators that adequate Russian funding will be provided soon, and the recent cabinet shake-up in Moscow, it is likely RSA elements will experience further delays.[27]

All of this delayed the launch of the American-built Unity module, a connecting node. The Russians were also late with the completion of the Zvezda service module, providing living quarters and life support systems. Accordingly, the first crew for the ISS could not occupy the station until November 2000, a delay of more than a year from the reoriented schedule made in the mid-1990s.

Throughout the remainder of the twentieth century, the sixteen countries on four continents engaged in building ISS hardware and software struggled to meet the assembly schedule. It would be easy to condemn their efforts as fruitless because of a seemingly insurmountable set of problems, schedule delays, and cost increases, but the consistent and oftentimes innovative solutions to overcome these challenges deserve some praise. To this day, the international partners have continuously readjusted their efforts, along with their levels of financial involvement and schedule commitments, and ultimate success remains likely. Further, modifications to the assembly sequence, ground operations, and on-orbit activities all require integration and various levels of coordination and joint approval. The U.S. developmental effort cannot be isolated from these occurrences and their associated impacts.[28]

Even as the program experienced cost growth and schedule slippage associated with this broad level of international involvement, there was excitement and apprehension when a Russian Proton rocket launched from the Baikonur Cosmodrome, Kazakhstan, and carried Zarya into orbit on November 20, 1998. It worked properly on arrival, but getting there had been almost too much to endure.[29]

On-Orbit Assembly to 2002

Beginning in 1994, the International Space Station project aimed toward the new direction mandated in the redesign that had taken place the year before. At that time, NASA operated under a constraint of $2.1 billion per fiscal year for ISS funding and with a total projected cost of $17.4 billion. An independent panel for Cost Assessment and Validation (under Jay Chabrow) concluded in 1998, "NASA's schedule and cost commitments were definitely success-oriented, especially considering the new realigned contracting approach with a single Prime contractor and that the specifics of Russia's involvement were just being definitized."[30] In other words, just as the Russians had been doing, NASA over-promised on what it could deliver for the money expended and the schedule agreed upon.

In addition, transitioning from the Space Station Freedom contracts to the ISS presented some challenges. As it brought Russia into the program, in March 1994 NASA undertook a full systems design review, thereby effectively completing the redesign but establishing a new baseline for the system. NASA and the partners devised an ISS assembly sequence on November 28, 1994, that reflected an initial occupancy in November 1997, with ISS assembly complete by June 2002. But these rosy forecasts could not mitigate the reality of the station's complexity. In April 1994 Canada shifted away from human spaceflight and space robotics toward space communications and Earth observation. As a consequence, NASA was forced to assume more responsibility (and about $200 million more in costs) for the extravehicular robotics function previously under the purview of the Canadian Space Agency.

The program also experienced some shifting in requirements. For example, in June 1994 the Centrifuge Accommodation Module (CAM), which had been a part of the Space Station Freedom design but was not identified specifically as integral to the ISS, reentered the program but without additional funding for it. The CAM was a much-valued laboratory that would house an 8.2-foot-diameter centrifuge, the essential component of a larger complement of research equipment dedicated to gravitational biology. It would make possible the use of centrifugal force to simulate gravity ranging from almost zero to twice that of Earth. It could also imitate Earth's gravity for comparison purposes while eliminating variables in experiments, and might even be used to simulate the gravity on the Moon or Mars for experiments that would provide information for future space travels. Although the scientific community applauded the addition, paying for the module proved yet another challenge. As a result, NASA later negotiated with Japan for the CAM in return for paying the launch costs of Japanese modules and astronauts.[31]

Even so, in 1999 NASA's international partners—Canada, the European Space Agency, Japan, and Russia—were expected to contribute the following key elements to the International Space Station, according to a NASA fact sheet:

- Canada is providing a 55-foot-long robotic arm to be used for assembly and maintenance tasks on the Space Station.
- The European Space Agency is building a pressurized laboratory to be launched on the Space Shuttle and logistics transport vehicles to be launched on the Ariane 5 launch vehicle.
- Japan is building a laboratory with an attached exposed exterior platform for experiments as well as logistics transport vehicles.
- Russia is providing two research modules; an early living quarters called the Service Module with its own life support and habitation systems; a science power platform of solar arrays that can supply about 20 kilowatts of electrical power; logistics transport vehicles; and Soyuz spacecraft for crew return and transfer.
- In addition, Brazil and Italy will be contributing some equipment to the station through agreements with the United States.[32]

The first two station components, the Zarya and Unity modules, were launched and joined together in orbit in late 1998. By that time, several others were nearing completion at factories and laboratories in the United States and elsewhere.

Orbital assembly of the International Space Station also began a new era of hands-on work in space, involving more spacewalks than ever before and a new generation of robotics. To complete it, the Space Shuttle and two types of Russian vehicles have been slated to launch a total of forty-six missions (thirty-seven Shuttle flights and nine Russian rockets). As of September 2002, thirteen Shuttle missions and three Russian launches have been completed. The first was the FGB or Zarya control module, placed in orbit in 1998.

Zarya was followed quickly by the Unity connecting module, taken aloft on Space Shuttle mission STS-88. Launched from the Kennedy Space Center on December 4, 1998, aboard the Endeavour, the Unity module is a six-sided connector for future station components. This was the first of the thirty-seven planned Shuttle flights to assemble the station. Endeavour's crew rendezvoused with the already orbiting Zarya module and attached it to Unity on December 6, 1998.

The crew then finished the connections during three spacewalks. They also entered the interior of Unity and Zarya to install communications equipment and complete other assembly work. Unity's six docking ports would provide access to ISS for all future U.S. modules. When the Shuttle left, Unity and Zarya were in an orbit 250 miles above Earth, monitored continuously by flight controllers in both Houston and Moscow.[33]

The next mission to ISS began on May 27, 1999, when the Shuttle Discovery performed the first docking with the International Space Station on May 29, 1999, as part of STS-96. Discovery's crew unloaded almost two tons of supplies and equipment for the station, including clothing, laptop computers, water, spare parts, and other essentials. They also performed one extravehicular activity to install a U.S.-developed spacewalkers' "crane," the base of a Russian-developed crane, and other spacewalking tools on the station's exterior to await use by future station assembly crews. Discovery fired its thrusters to reboost the station's orbit, undocked on June 3, 1999, and landed back on Earth three days later.[34]

Although the launches of Zarya and Unity had been good news, and some work had taken place after their assembly in December 1998, a nineteen-month hiatus followed before the Russians completed the Zvezda. This module provided living quarters, life support, navigation, propulsion, communications, and other functions for the early station. Without it, crews could not remain on the ISS unless the Space Shuttle was present. The Zvezda was the first fully Russian contribution and the core of the Russian station segment. Launched without a crew on July 25, 2000, the module docked with the orbiting Zarya and Unity by remote control, and the gates were opened for ISS assembly. ISS guidance and propulsion systems took over those functions from Zarya, which then became a passageway between the Unity and Zvezda modules.[35]

According to the original assembly sequence for the redesigned space station in 1993, the service module was to have been in orbit in April 1998. However, the series of delays previously discussed pushed the launch date back by some two years, then even further when the Proton rocket fleet suffered two launch failures. Eventually these were traced to manufacturing problems within upper stage rocket engines, and the system had to be redesigned. Meanwhile, as we have seen, NASA developed contingency plans that were also plagued by delays and cost overruns. But in the end, the launch of Zvezda paved the way for the launch of the U.S. laboratory module Destiny, and occupation by a permanent crew.[36]

In anticipation of the first crew boarding the ISS, the Space Shuttle Atlantis (STS-106) made a second cargo flight beginning on September 8, 2000. Atlantis docked with the ISS carrying supplies to be transferred to the interior. The Shuttle's crew performed a spacewalk to attach a telescoping boom to the Russian spacewalker's crane, left on the station's exterior during mission STS-96, as well as to conduct other assembly tasks. After Atlantis undocked, ISS remained unoccupied until October when, prior to the Expedition 1 crew's arrival, STS-92 delivered the

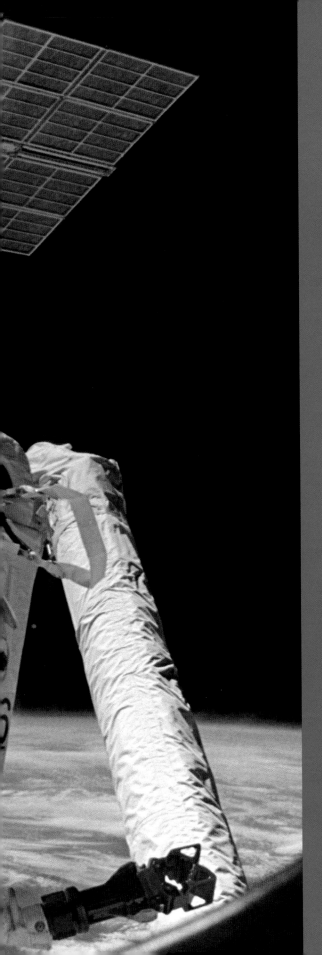

STS-88 mission specialist James Newman, holding on to a handrail, waves during the first of three EVAs that were performed . The orbiter can be seen reflected in his visor. In December 1988 these EVAs assembled on orbit the Unity and Zarya components of the ISS. (NASA photo, no. STS088-343-025)

Backdropped against white clouds and blue ocean, the ISS moves away from the Space Shuttle Discovery on June 3, 1999. The U.S-built Unity node *(top)* and the Russian-built Zarya or FGB module (with the solar array panels deployed) were joined in a December 1998 mission. A portion of the work that astronauts Tamara E. Jernigan and Daniel T. Barry performed during their May 30, 1999, spacewalk is evident at various points. For example, they installed the Russian-built crane, called Strela. (NASA photo, no. STS096-333-021)

The Zvezda service module is shown here under construction at the Krunichev State Research and Production Space Center (KhSC), Moscow, in September 1997. In the foreground is the forward portion of the module, including the spherical transfer compartment and its three docking ports, one facing up, one down, and one forward. The forward port will dock with the connected FGB and Node 1 following the service module's planned launch on a Russian rocket in December 1998. Zvezda will serve as the early cornerstone for human habitation of the station, providing preliminary living quarters, life support, guidance, navigation, and propulsion. (NASA photo, no. S97-12607)

Z-1 Truss, Pressurized Mating Adapter 3, and four Control Moment Gyros. Astronauts performed four days of spacewalks to finish these connections.[37]

On October 31, 2000, the first crew to occupy the International Space Station inaugurated a new era in space history. When American astronaut Bill Shepherd and Russian cosmonauts Yuri Gidzenko and Sergei Krikalev lifted off in a Russian Soyuz spacecraft from the Baikonur Cosmodrome, bound for their new home aboard the ISS, it represented the last day on which there were intended to be no human beings in space. Shepherd (of Babylon, New York) was commander of the three-person Expedition 1 crew, the first of several who would live aboard the space station for periods of about four months. As new elements were added to the orbiting outpost, Shepherd, Gidzenko, and Krikalev worked on assembly tasks and conducted early science experiments.[38]

STS-97 became the last Space Shuttle mission of the twentieth century, and it proved an exceptionally important one. This time the Space Shuttle Endeavour and its five-member crew installed the first set of U.S. solar arrays and became the first crew to visit Expedition 1. The arrays set the stage for the launching of the Destiny Laboratory Module, which arrived in February 2001 on STS-98. The five STS-98 astronauts also relocated Pressurized Mating Adapter 2 from the end of Unity to the end of Destiny, in preparation for future missions. The U.S.-built Destiny module, the first station laboratory, promised a center place for future research activity on the ISS. Discovery used its robotic arm to maneuver the new laboratory into position, then the crew performed three spacewalks to finish the installation.[39] As planned, Destiny quickly became the most important science component of the ISS, with its internal interfaces to accommodate the resource requirements of twenty-four equipment racks. Thirteen of these were International Standard Payload Racks (ISPR), and the rest were for such uses as controlling ISS systems.[40]

In March 2001 the first crew rotation flight to ISS arrived. STS-102 delivered the Expedition 2 astronauts and returned Expedition 1 to Earth after four and a half months in space. STS-102 also attached to the station the first Multi-Purpose Logistics Module, Leonardo, unloaded it, and

This is how one artist believed the ISS would appear after the crew of STS-97 deployed the P6 Integrated Truss Segment. The truss holds the U.S. solar arrays and a power distribution system. (NASA photo, no. JSC2000-E-29786)

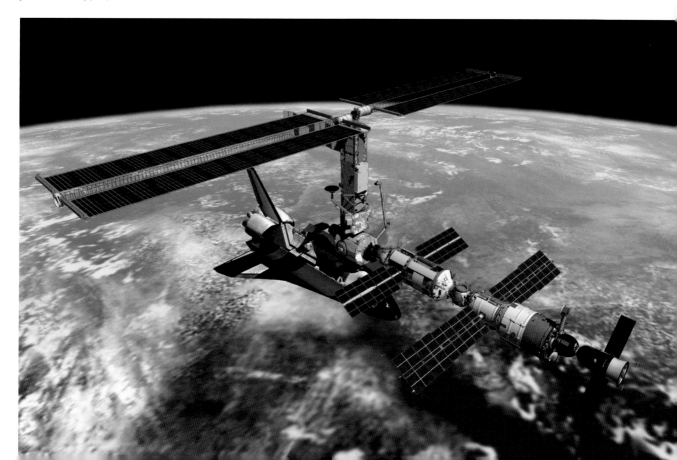

returned it to Earth. Logistics modules were built by the Italian Space Agency as reusable cargo carriers. One month later STS-100 delivered the station's robot arm, also known as the Space Station Remote Manipulator System, and the Raffaello Multi-Purpose Logistics Module. The delivery of the arm allowed for the arrival of the joint airlock, which was installed during STS-104's visit in July 2001.[41]

NASA accomplished a number of missions to the ISS during the remainder of 2001. The next Shuttle mission was STS-105, in mid-August 2001, which delivered the Expedition 3 crew and returned Expedition 2 to Earth. The Leonardo Multi-Purpose Logistics Module made its second trip to the station during STS-105. ISS expansion continued with the arrival of the Russian Docking Compartment, Pirs, on September 16, 2001. The next flight to visit the space station was STS-108, which arrived early that December to deliver the Expedition 4 crew and return Expedition 3 to Earth. The first mission to visit the station in 2002 was STS-110, which installed the first piece of a nine-segment truss to hold modules and solar arrays. In mid-October 2002, STS-111 delivered the Expedition 5 crew along with supplies and equipment, and returned the previous crew.[42]

As it entered 2002, the ISS assembly effort continued to be hampered by some of the same problems it had faced in preceding years. Although reasonable progress was maintained on U.S. elements, many challenges resulted in increased cost and schedule erosion. U.S. development problems continued to be overshadowed by Russian funding shortfalls and delays in its commitments; however, U.S. production delays and the incorporation of much-needed multi-element integrated testing also slowed the process. Recurring problems took their toll, including late part and component deliveries.[43]

HOW TO GET HOME SAFELY

When the space station is completed, an international crew of up to seven should be able to live and work in space for between three and six months. The major problem in sending up a crew of seven at this point is returning them safely to Earth in the event of an emergency. This led throughout 2001 and early 2002 to an enormous debate over the value of the space station, as funding and schedule again came under fire from critics. To ensure the crew size of seven, in 1994 NASA had added to the program a U.S.-developed crew return vehicle (CRV) without requesting additional funding to pay for it. NASA carried the CRV as an overage in the budget for ISS until funding was finally allocated, in the fiscal year 1999 budget submission to Congress. But that was only a small part of the X-38 crew return vehicle's problems.[44]

As always, NASA began the project with optimism. The agency had a record of excellence that had been forged during Project Apollo in the 1960s, and success after success had followed, including landings on Mars and voyages to the outer planets of the solar system. A culture of

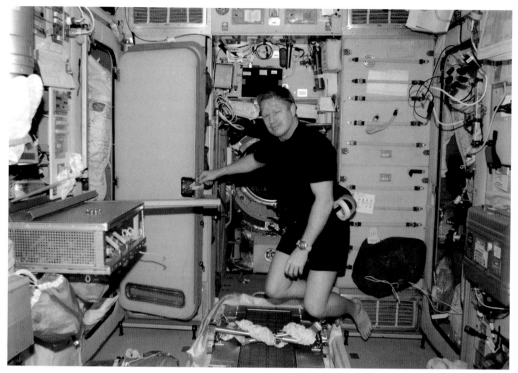

Astronaut William M. Shepherd, commander for the Expedition One mission, floats in the microgravity, shirtsleeve environment of the ISS's Zvezda service module in December 2000. Shepherd was taking a break to install various furnishings. The image was taken with a digital still camera and downlinked from the station to ground controllers in Houston. (NASA photo, no. ISS01-E-5128)

competence permeated NASA. Actor Robert Guillaume voiced the general public's belief when he said in the popular sitcom *Sports Night,* "You put an X anyplace in the solar system, and the engineers at NASA can land a spacecraft on it."[45] Unfortunately in the case of the X-38, cost and schedule combined to defeat the program.

It began at Dryden Flight Research Center in 1992, when Dale Reed, a veteran of the "lifting body" research of the 1960s and 1970s, started work for the Johnson Space Center to test a CRV concept modeled on earlier efforts. This proved successful, and in-house development of the X-38 began at JSC in early 1995. In the summer of 1995, early flight tests were conducted by dropping from an aircraft designs for the CRV at the Army's Yuma Proving Ground in Arizona. In early 1996 NASA awarded a contract to Scaled Composites of Mojave, California, for the construction of three full-scale atmospheric test airframes. The first was delivered to JSC in September 1996, where it was outfitted with avionics, computer systems, and other hardware in preparation for flight tests at Dryden. The second vehicle was delivered that December, and testing progressed to the verge of an unpiloted spaceflight in late 2000. About one hundred people quickly went to work on the project at Dryden and JSC.[46]

The X-38 CRV, when operational, was intended as the first reusable human spacecraft to be built in more than two decades. It was designed to fit the unique needs of a space station "lifeboat"—long-term, maintenance-free reliability always in turnkey condition—ready to provide a crew of up to seven a quick, safe trip home under any circumstance. The design of the X-38 incorporated the wingless lifting body concept pioneered at Dryden Flight Research between 1966 and 1975.[47] Data from those earlier aerodynamic studies had contributed to the design and operational profile of the Space Shuttle and then reemerged in the CRV program. The X-38 would be carried to the space station in the cargo bay of a Shuttle and attached to a docking port. If an emergency arose, the CRV could be boarded, undocked, and, after a deorbit engine burn, would return to Earth much like a Space Shuttle.

Not a true spacecraft in the traditional sense, the vehicle's life support system could sustain a crew for about seven hours. Once it reached the denser atmosphere at about 40,000 feet, a steerable parafoil would be deployed to carry it through the final descent and landing. It would also be fully autonomous, in case the crew was incapacitated or otherwise unable to fly it; it could return to the landing site using onboard navigation and flight control systems. Backup systems would, of course, allow the crew to pick a different site and steer the parafoil to a landing.[48]

The design closely resembled that of the X-24A lifting body flown at Dryden between 1969 and 1971. That vehicle, like the CRV, generated aerodynamic lift—essential to flight in the atmosphere—from the shape of its body. The twenty-eight research missions flown by the X-24A helped demonstrate that hypersonic vehicles returning from orbital flight could be landed on conventional runways without power. The three prototype X-38s used in the atmospheric flight testing program were 24.5 feet long, 11.6 feet wide, and 8.4 feet high, approximately 80 percent of the planned size of the CRV. They were designated V131, V132, and V131R. The first of these was modified for additional testing beginning in the summer of 2000, and thereafter it carried the designation V131R. A fourth prototype that has not yet been built, V133, is intended to incorporate the exact shape and size of the planned CRV. These vehicles had shells made of composite materials such as fiberglass and graphite epoxy and were strengthened with steel and aluminum at stress points. Weighing between 15,000 and 25,000 pounds, they landed on skids—reminiscent of the famous X-15 research aircraft of the 1960s—instead of wheels.[49]

Also not built was a fully space-rated X-38 CRV prototype, numbered V201, to test the concept under operational conditions. Its inner compartment, representing the crew area, would be

a pressurized aluminum chamber. A composite fuselage structure would enclose the chamber, and the exterior surfaces would be covered with a Thermal Protection System (TPS) to withstand the heat generated by air friction as the vehicle returned to Earth through the atmosphere. The TPS would be similar to materials used on the Space Shuttle, but much more durable: carbon and metallic-silica tiles for the hottest regions, and flexible blanket-like material for areas receiving less heat during atmospheric reentry.[50]

Much of the technology being used in the X-38 was to be off-the-shelf, to avoid lengthy research and development efforts. For instance, the flight control computer and software operating system had been commercially developed and already used in many aerospace applications. Inertial navigation and global positioning systems, similar to units used on aircraft throughout the world, would be linked to the flight control system to steer the vehicle along the correct reentry path. Finally, the U.S. Army originally developed the design of the parafoil used to land the X-38.[51]

Even though the CRV used proven technology for most of its systems, it quickly became mired in cost and schedule problems. The technical complexity of the development task, the international character of the effort, and the time phase for completing it all conspired to push costs beyond acceptable bounds. Original estimates to build a capsule-type CRV amounted to more than $2 billion in total development cost. After paring, the X-38 program team arrived at an estimate of $1.3 billion. A review of the ISS program costs in the fall of 2001 noted that for the CRV and other elements, "There is inadequate current costing information associated with the non-U.S. components." It also found that "Project interruptions will have cost impact on all of the elements under consideration."[52]

By the end of 2001 the CRV had been eliminated, not so much because of its cost overruns as because of huge overruns throughout the ISS program. For those looking for places to trim the ISS budget, the CRV proved an easy target. Hints of this came early. NASA administrator Daniel S. Goldin testified to Congress in April 2001:

> The U.S. CRV has a significant set of design activities to accomplish before we are ready to enter into a production contract. Just last year, NASA's Integrated Action Team, focusing on program management excellence, concluded that technology risk reduction programs and design definition must be concluded before committing to production contracts to best insure that cost, schedule and technical targets can be realized. Given the magnitude of planned funding dedicated to CRV and the remaining definition work, funding allocated

The X-38 descends during a July 1999 test flight at the Dryden Flight Research Center, California. This was the fourth free flight of the test vehicles in the X-38 program and the second free flight test of Vehicle 132, or Ship 2. Using available technology and off-the-shelf equipment to decrease development costs, the X-38 research project was designed to develop the technology for a prototype emergency crew return vehicle (CRV), or lifeboat, for the ISS. (NASA photo, no. EC99-45080-21)

The Space Shuttle Discovery approaches the ISS during the STS-105 mission, August 2001. Aboard Discovery are the crew members of ISS Expedition Three: Frank L. Culbertson Jr., mission commander; and cosmonauts Vladimir N. Dezhurov and Mikhail Tyurin, flight engineers. They replaced the Expedition Two crew that had been living on the ISS for the five months since March 2001. Visible in the payload bay of Discovery are the Multi-Purpose Logistics Module Leonardo, which stores supplies and experiments to be transferred into the ISS; the Integrated Cargo Carrier, which carries the Early Ammonia Servicer; and two materials experiment containers. (NASA photo, no. ISS02-E-9744)

for the CRV production phase has been redirected to help resolve ISS core content budgetary shortfalls. NASA has initiated discussions with the European Space Agency (ESA) on a role in the CRV project. Critical efforts such as X-38 atmospheric flight testing and some preliminary CRV design work and linkages with CTV [crew transportation vehicle] under SLI [Space Launch Initiative] will continue so as to maintain viable options for future CRV development. The planned space flight test of the X-38 is under review as part of the program assessment.[53]

The new George W. Bush presidential administration directed NASA to halt further development of the CRV, and while NASA sought to find ways to secure a CRV without direct agency funding, the results were disappointing.[54]

ISS Cost Growth: A Tiger by the Tail

When the Freedom program became the International Space Station, NASA believed it could build the station on a budget of $17.4 billion over a ten-year period. It could not have been more wrong had it set out to offer disinformation. After three years of insisting that it could meet its initial estimate, in September 1997 NASA finally conceded that it could not. Cost overruns on Boeing's contract and the need for an additional $430 million for NASA in 1998 were announced. NASA began transferring funds from its other programs into space station construction. By the time of a major review in fall 2001, the estimated U.S. portion of the ISS development stood at about $23 billion.[55]

Following the cost growth announced in late 1997, NASA convened an independent Cost Assessment and Validation team headed by Jay Chabrow. The Chabrow committee concluded that the station could cost up to $24.7 billion and take ten to thirty-eight months longer to build than NASA believed. NASA did not accept those findings in their entirety, but agreed that it would take longer to build and would cost $1.4 billion more, bringing the ISS estimate at that time to $22.7 billion. To help control costs, Congress placed a cap on the station limiting expenses to $25 billion, with a $2.1 billion annual maximum. In 2001 another review team found, "The financial focus is on fiscal year budget management rather than on total Program cost management."[56]

Most important, until after the 2000 presidential election NASA managers for the ISS kept silent the fact that they were tracking about $4 billion in cost growth for fiscal years 2002 to 2006. This not only blew the legislated cap for the total cost, but it also demolished the annual funding constraints. Since 1998 the estimated total expense of building the International Space Station has grown every year by an average of $3.2 billion. In 2001 the complete ISS was estimated at more than $30 billion, with a conclusion date of 2006. Much of the U.S. hardware had already been built, and more was in the process of being finished. Some had been assembled

Astronaut Nancy J. Currie *(left)* and cosmonaut Sergei K. Krikalev *(right),* both mission specialists, use rechargeable power tools to manipulate nuts and bolts on the Zarya module. Astronaut Robert D. Cabana *(in rear),* mission commander, moves along the rail network in the background. The six STS-88 crew members had earlier entered through the Unity connecting module. Rails, straps, and tools indicate that the crew members had already been working for some time when this photo was taken. (NASA photo, no. STS088-359-037)

on orbit and made operational. Many at NASA assumed by this time that barring a catastrophe, the program was past its major pricetag hurdles.

Instead, NASA leaders found themselves with a set of additional costs that had to be met to complete the project. Most of these had been pushed forward in time over the years and were now coming due. The CRV, habitat module, propulsion module, and the bulk of the science program intended for ISS all fell under the budget ax. An air of crisis permeated the program all through the summer and fall of 2001.[57]

In June 2001 NASA leaders struck an agreement with the incoming Bush administration regarding adjustments in content and funding to partially address the nearly $1 billion shortfall for ISS that remained, even after other reductions had been made. This effort pushed the funding profile out to fiscal year 2003, with a deficit of $484 million remaining for 2004 to 2006.

A subtext to this agreement was that NASA had to reorganize the ISS and demonstrate that it could manage a complex large-scale project such as this, and, after two years of probation, in 2004 the agency might be granted some additional funding to complete the space station.[58]

NASA and the White House thus agreed to a plan that proved exceptionally difficult to realize. NASA's strategy was to issue a "management challenge" to find cost savings that would make up for the shortage. The Bush administration directed NASA to form an independent external task force to assess credibility of this plan. That report, issued in November 2001, set off a firestorm of NASA "bashing" in the media. The *New York Times* editorialized:

> The international space station—the centerpiece of the American manned space program—has encountered severe cost and management difficulties that could damage its value as a platform for conducting "world class" science, the supposed goal of the project. No more urgent task will confront the next administrator of the National Aeronautics and Space Administration than to get this troubled program under control. The station, which is now partially assembled and circling overhead with three astronauts aboard, has had a difficult evolution since President Ronald Reagan first ordered it built in 1984. Costs have escalated, forcing repeated delays, redesigns and downsizings, and Congress has considered no less than 25 different proposals to terminate the program.
>
> This page has long been skeptical of the value of the station given its rudimentary technology, huge cost and lack of a clear mission that required the permanent presence of astronauts on a platform in low earth orbit. . . .
>
> Nobody doubts that the station has been a great engineering achievement, demonstrating that astronauts can indeed erect a platform in space with few glitches. It is NASA's managerial performance that has been dismal. Early this year the agency shocked the administration and Congress by revealing that it would overshoot a $25 billion Congressional cap by almost $5 billion, exclusive of the $18 billion needed for Shuttle flights to boost astronauts and hardware into orbit. Worse yet, nobody could be sure that that was the end of the cost overruns.

The *Times* added that although the Bush administration had forced NASA to scale back its plans for the station, it should also hold the agency responsible for proving itself. "Should NASA fail to meet the challenge over the next two years," wrote the editorialist, "we could be left with a station of minimal capabilities." It was not a very satisfactory situation.[59]

The administration's task force, under the direction of former NASA and Martin Marietta senior executive A. Thomas Young, brought forward a set of scathing findings. The currently planned program was "not credible," they concluded, pointing to deficiencies in management structure, institutional culture, cost estimating, and program control. They expressed concern over the validity of the original $8.3 billion cost estimate in the NASA–Bush administration agreement and the probability of achieving all the projected savings. The task force recommended an overhauling of program management, clearly defining the science goals, and significantly reducing workforce levels to keep the currently planned three-person on-orbit capability within the projected budget. In addition, they said, the Space Shuttle flight rate could be reduced to four per year to save funds to offset increased ISS costs. A performance-based approach was recommended, whereby NASA would demonstrate credibility over a sustained period as a prerequisite to proceeding beyond the three-astronaut capability. The task force cautioned that it would be very difficult for NASA's culture to change to the degree required to make the program succeed.[60]

During the review, Young's commission identified new cost increases totaling $366 million. Several more concerns were identified, but not costed, in the areas of contractor rates, international partner costs stemming from scaling back, risks of research, and inadequate planning for necessary system upgrades. Offsets in the amount of $440 million from within the program were identified, primarily from staff reductions and reevaluation of sustaining engineering needs. The task force found approximately $1 billion of potential additional savings from within the human spaceflight account ($668 million from reduced Space Shuttle flights and $350 to $450 million from an ongoing NASA Strategic Resources Review savings effort).[61]

In the midst of this crisis, Dan Goldin resigned as NASA administrator after nearly ten years on the job. The new presidential administration had made it clear that he would not be retained, and in November 2001 he took the opportunity to retire. This meant that a new cast of leaders would have to clean up the ISS budget woes. George W. Bush sent Sean O'Keefe to NASA, and he took office just as 2002 dawned.

O'Keefe had a reputation as a firm manager, having served previously as comptroller of the Department of Defense and as Secretary of the Navy under the earlier Bush administration of 1989–93. O'Keefe outlined three basic themes as the basis for how he intended to approach implementation of the program. He said, "The dominant issue on the ISS is to get the core complete baseline request right so as to begin with an integrated mission established as the fundamental baseline." This would involve pursuing "five specific objectives," he said:

1. Prioritize the science and technology agenda for the ISS.
2. Analyze the remaining engineering challenges. The ISS is largely operational now, but there are still steps required to complete it.

3. [Assess] Financial issues that surround estimating for cost complete. To do that objective is to capture the total cost over what is in place now. An independent assessment as well as internal assessment will be required.

4. Take a look at the international commitments for the ISS and the effect that new ISS plans have on this.

5. Understand all other elements that must be considered to support this project—Space Shuttle operations being the most prominent.

O'Keefe's second theme involved thinking of projects such as the ISS as integrated facilities—integrated across the agency. It remains to be seen how successful he will be in tackling these important challenges.[62]

By the first part of the twenty-first century, the International Space Station program had enjoyed enormous success and suffered from serious questions about mismanagement and cost overruns. As Thomas Young's commission ironically concluded, "The management focus is on technical excellence and crew safety with emphasis on near-term schedules, rather than total program costs."[63] It was that emphasis on technical excellence and safety that caused such a flurry of criticism. The commission also noted: "The institutional needs of the Centers are driving the Program, rather than the Program requirements being served by the Centers. The impact of institutional management is clearly indicated in the overall staffing levels of the program. The institution, not the program, controls the majority of these resources and timely destaffing is significantly hindered."[64]

Failure to take cost into consideration represented a longstanding problem at NASA. One of the fundamental tenets of program management has been that three critical factors—cost, schedule, and reliability—are interrelated and must be managed as a group. Many also recognized these factors' constancy; if managers held cost to a specific level, then one of the other two, or both to a lesser degree, would be adversely affected. This problem has held true for the ISS program, just as it did previously. Since humans are flying on the ISS, the program managers placed a heavy emphasis on reliability to ensure crew safety. The schedule proved more flexible, but slipping the launch dates affected other issues as well. The significance of both of these factors forced the cost factor much higher than might have been the case with a more leisurely ISS program.[65]

As the ISS continues to develop, with both great successes and continuing challenges, the budget will dictate what direction the final program takes. The funding issues associated with ISS will affect future years, primarily in the 2004 to 2006 period. Resolution will be addressed through several steps:

• Removing the CRV, Habitation Module and related ECLSS [environmentally closed life support system] development, Propulsion Module, and integrating Node 3 and the Centrifuge Accoummodation Module, as well as realigning some research facilities to better fit on-orbit capabilities

- Examining alternative means of providing habitation, crew return capabilities, and on-orbit research
- Increasing headquarters insight and involvement in additions or changes to program content
- Implementing measures to improve cost estimating and projection accuracy
- Conducting an independent external review of ISS budget projections in order to validate supporting assumptions, impact of planned cost mitigation actions, and identification of further actions needed to meet long-range objectives

NASA leaders voiced a strong commitment to supporting and responding to the fiscal constraints on ISS and to identifying the steps necessary to achieve the overall objectives of the program within the funding limitations and guidelines prescribed by the Bush administration and Congress. Without a doubt, more debate and decision-making will be necessary before the ISS is completed.[66]

In the classic science fiction movie released in 1968, *2001: A Space Odyssey*, director Stanley Kubrick depicted a bold future in space, with a station central to human exploration. With the operational activities of ISS in 2001 and 2002, it is appropriate to reflect on the "space odyssey" of reality in relation to Kubrick's powerful vision. First, and perhaps most important, the Cold War is over, and the United States and former Soviet Union have joined with others to make a reality the long-held vision of a space station in Earth orbit. With the arrival of the first crew on ISS in 2000, residency in space has become the norm, rather than the exception, for the spacefaring nations of the world. Accordingly, at the dawn of this new millennium, humanity has accomplished much in space and may be closer to the vision set forth in *2001: A Space Odyssey* than many people think. Humans have a space station and a reusable Shuttle—and both capabilities will be upgraded in the future—and they have been to the Moon and may well soon return. Computer technology surpasses that depicted in the film. Probes have been sent to every planet of the solar system except Pluto, and they are finding fascinating hints about the possibility of life elsewhere in the universe.

Most significantly, space exploration is being undertaken mostly as a cooperative venture of many nations. Absent the need to expend vast resources for an arsenal of nuclear weapons, a future built on peaceful utilization of space is our most important opportunity for the future. Indeed the odyssey continues, and the International Space Station is a centerpiece of it. It is perhaps somewhat different from that depicted in Kubrick's film, but it is every bit as enticing.

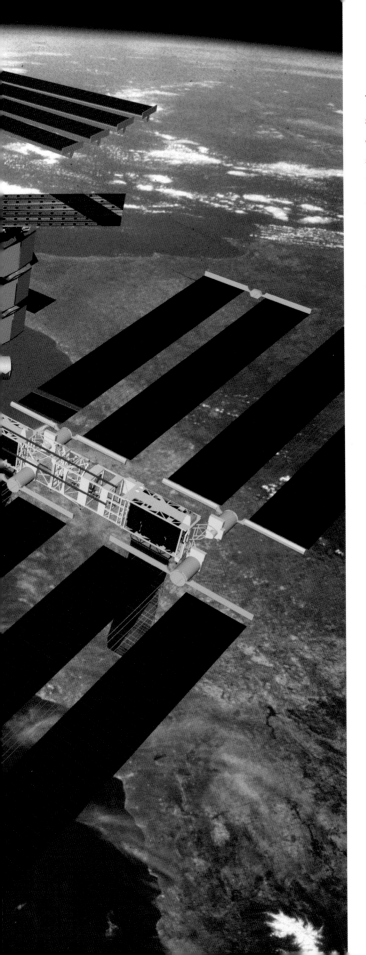

This 1997 digital artist's impression shows the ISS passing above the Straits of Gibraltar and the Mediterranean Sea after all assembly is completed. By that time the station would be powered by almost an acre of solar panels and have a mass of almost 1 million pounds. Its pressurized volume would be roughly equivalent to the space inside two jumbo jets. The first piece of the station was launched in November 1998, beginning a challenging five-year, forty-five-flight sequence of assembly in orbit. The first crew went aboard ISS in October 2000. (NASA photo, no. S97-10538)

CHAPTER 7

Epilogue

Whither the
International Space Station

The goal of a permanent presence in space has driven space exploration since at least the beginning of the twentieth century, and the ISS has been accepted now as the current pièce de résistance of human spaceflight. But does this scenario make sense for the future? Will the ISS open the space frontier, as so many advocates believe? Will it enable humanity's movement beyond this planet, through its contribution to knowledge and its fostering of key technologies? This epilogue explores such questions and offers comments on the spacefaring enterprise in the first part of the twenty-first century.

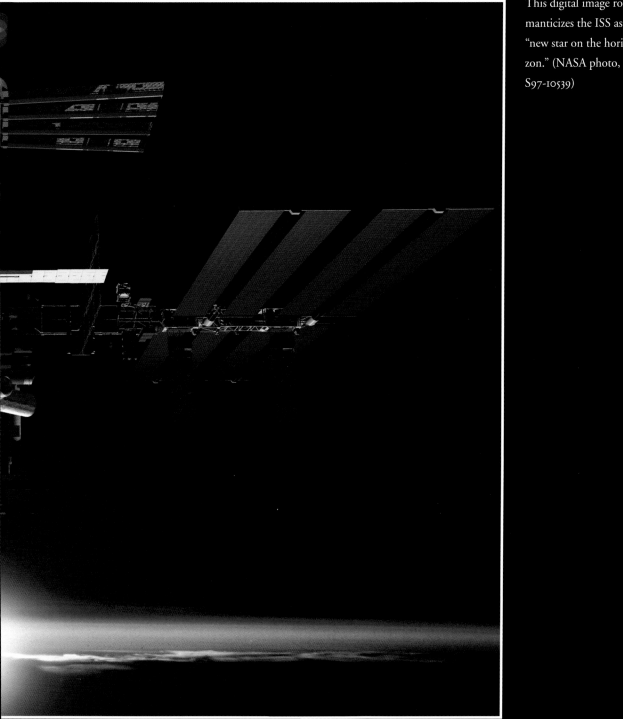

The Challenge of Assembly on Orbit

Assembly of the International Space Station on orbit has presented far more challenges than ever envisioned by Wernher von Braun and his cohorts of an earlier era. Astronauts in specially developed and hard-to-maneuver space suits, using specially developed and hard-to-maneuver tools, have been performing tasks that, although simple on Earth, are rigorous, time consuming, and fatiguing in the harsh microgravity environment of low Earth orbit. Anchoring themselves to supports and rigs, astronauts laboriously use hand and power tools to bolt the pieces of the ISS together.[1]

To assemble the million-pound ISS, low Earth orbit has been a day-to-day construction site since 1998, and the work is not projected to be completed until 2006. By far the most complex activity asked of astronauts is performing spacewalks to assemble the ISS. They are required to handle, in completing the project, more than 100 different components launched on about 46 space flights—using 3 different types of rockets—and then bolt, latch, wire, plumb, and fasten them together.

Because of its unprecedented complexity, everyone involved in the ISS project expects to encounter numerous surprises and challenges during the orbital construction work. But to prepare for them, engineers and astronauts methodically practice procedures, prepare tools, test equipment, and build experience on the ground using simulators and underway "neutral bouyancy tanks." About 160 spacewalks, totaling 1,920 work hours, will be performed to assemble and maintain the station. By contrast, since astronaut Edward G. White first stepped out of the orbiting Gemini II spacecraft in 1965 to become the first American to walk in space, NASA astronauts have completed only about 377 hours of spacewalks (Ed White died in the Apollo 1 fire on January 27, 1967).[2]

During the first few assembly missions, there was no U.S. capability for spacewalks from the station without the Space Shuttle being present, since ISS had no airlock at that time. The Zvezda service module provided a capability for station-based Russian spacewalks, but only using Russian space suits. The U.S. capability was not available until the joint airlock module was attached to the station during the ninth Space Shuttle assembly mission, STS-104, in July 2001. It can be used by both Russian and U.S. space suit designs and allows "pre-breathing" by spacewalkers at 10.2 pounds per square inch (psi) to purge nitrogen from their bodies and thereby prevent decompression sickness, commonly called the bends, when they go to the 4.3 psi pure oxygen atmosphere of a space suit. To help with the process, the vehicle's cabin pressure is lowered to 10.2 psi at least a day ahead of the EVA.[3] As a result, astronauts cannot simply put on a space suit and venture outside the ISS. This is a severely limiting situation, and the ISS community has a major challenge in developing new space suits that will enable more ready use.

To build and maintain the International Space Station, spacewalking astronauts have been

Astronaut Jerry L. Ross, STS-88 mission specialist, is pictured during one of three spacewalks that were conducted on the twelve-day mission between December 4 and 15, 1998. Perched on the end of Endeavour's Remote Manipulator System arm, astronaut and mission specialist James H. Newman recorded this image. Newman can be seen reflected in Ross's helmet visor. The solar array panel for Zarya can be seen along the right edge. This was the first of about 160 spacewalks totaling 1,920 work hours that will be required to complete the International Space Station. (NASA photo, no. STS088-355-015)

working in partnership with a new generation of robotic surrogates. The Shuttle's Canadian-built mechanical arm and a new space station arm operate both as cranes, to maneuver large modules and components, and also as space "cherry pickers," to maneuver astronauts to work areas. The Shuttle arm has been enhanced by a new Space Vision System (SVS) that helps the operator see around corners. The SVS employs video imagery and a series of markings on the objects being maneuvered to develop a graphical laptop computer display, allowing great precision even when visibility is obstructed. The intent is to allow astronauts to safely move objects with precision. The system was used operationally for the first time during the first assembly mission in 1998, as astronaut Nancy Currie, with her view partially obstructed, attached the first station component, the Zarya control module, to the second component, the Unity module. The arm, which represents the state of the art in robotics, has the capability to be latched to virtually any module of the station.[4]

There is much more to do before the ISS is complete, but the success thus far has been impressive in safely carrying out the assembly. At a fundamental level, the fact that thousands of components have been assembled into modules in space—and that they actually work properly—is astounding.

Science on the ISS

Many of those involved in building the International Space Station believe that its emphasis should be on the scientific return that will come from having in permanent Earth orbit a world-class research institution. Soon after the launch of its first elements in 1998, the ISS began to afford scientists, engineers, and entrepreneurs an unprecedented platform on which to perform complex, long-duration, and replicable experiments in the unique environment of space. Research opportunities for the future involved prolonged exposure to microgravity, and the presence of human experimenters in the research process. While an international crew of astronauts lives and works in space, the ISS may expand here on Earth as researchers use the technologies of "telescience" to control and manipulate experiments from the ground. Advancing communications and information technologies could allow Earth-bound investigators to enjoy a "virtual presence" on board the ISS along with an international cadre of "real time" astronauts and researchers, all taking their place in a world community that will use and benefit from an orbiting laboratory.[5]

Scientific inquiry represents the fundamental activity that can take place on the International Space Station at this point in time. NASA has soft-pedaled the idea of using ISS as a jumping-off point for future exploration of the Moon and planets, although that remains a long-term objective. Some have referred to the station as an "NIH [National Institutes of Health] in space," from which all manner of biotechnical discoveries might spring with important applications for Earth's

Entrepreneurs from Spacehab, Inc., may start the process of turning the ISS into a research park in space with their Enterprise science module. This facility will provide a microgravity research capability on the station using a fee-for-service system. (Courtesy of Spacehab, Inc.)

population. Others have emphasized the station's significance as a laboratory for the physical sciences, with materials processing in microgravity as the chief research, enhanced by work on orbit. Still others suggest that human-factor research will gain a leap forward because of the work on ISS, simply because data about changes to the bodies of astronauts engaging in long-duration spaceflight will expand the base of scientific knowledge. Finally, the ISS offers a platform for greater scientific understanding of the universe, especially about the origins and evolution of the Sun and the planets of this solar system. Those four endeavors—biotech research, materials science, human factors, and space science—represent the panoply of scientific opportunities offered by the ISS.

Do those activities fully justify the enormous cost of development, construction, and operation of the ISS? The answer depends very much on individual perspective. Generally speaking, space advocates argue for the ISS as a great boon to scientific knowledge, but the larger community has many questions as to whether the funds might have been more effectively applied to other equally or more valuable research projects. The *New York Times* editorialized: "In truth, it has never been clear just what science needs to be done on a permanently manned platform in space as opposed to an unmanned platform or an earthbound facility."[6] NASA's failure to define a clearly supportable science mission left some wits to suggest that ISS was very much like the movie *Field of Dreams,* in the which the central character plowed under his cornfield to erect a baseball diamond in Iowa because a voice said, "If you build it, they will come."[7] Just because the ISS was in orbit, they noted, why would anyone think it a foregone conclusion that the science program offered genuine value? The *New York Times* editor opined that if the station did not yield a rich scientific harvest within a couple of years, "NASA might consider downgrading [it] to an unmanned platform for automated experiments that would be tended occasionally by visiting astronauts.[8]

Even advocates of the space station raised concerns about the lack of clarity of a science mission. Representative Ralph M. Hall (Democrat, Texas), speaking before a gathering of senior aerospace officials on March 27, 2001, commented that the program was in "a time of transition," and that he and his colleagues in Congress had lost patience with both NASA and the George W. Bush administration's inability to resolve the problems of the ISS and define its science mission. "First, we need to decide what we are going to do about the reported cost growth in the Station program," he said. "And second, we need to give some serious thought to how we should operate and utilize the Space Station once it is fully operational. While the first challenge is the more immediate one, the second will be critical in determining whether the Station lives up to its potential in an affordable manner." While solving the funding concerns, he added, the United States had to commit to a robust space science program across a range of disciplines that would yield important results. "After all of the taxpayer dollars that have been invested in the Space Station," Hall said, "we will need to ensure that we wind up with the world-class research facility that we have been promised." As an aside to his prepared remarks, Representative Hall commented that once the ISS became operational, NASA had better find a way to use it effectively. He warned that some astounding scientific discovery should be forthcoming—a cure for cancer, specifically—or else the program could rapidly lose political support.[9]

The best answer perhaps came in late 2001 from a review directed by former NASA and Lockheed executive A. Thomas Young. It recommended that NASA concentrate on using the ISS for research on "engineering required to support humans in long-duration space flight."[10] That would help both to focus and to bound the NASA effort, as well as emphasizing scientific activities that would be needed when humanity ventured into deep space.

Accordingly, at present the critical component for research on the space station involves learning about how humans react to long-duration stays in a microgravity environment. Coining the term *bioastronautics* to cover this research agenda, NASA implemented beginning with the Expedition 3 crew a research program that sought to identify and characterize health, environmental, and other operational human biomedical risks associated with living in space. It also aimed to develop strategies, protocols, and technologies that might prevent or reduce those risks. Only by understanding exactly how the human body—the heart, lungs, muscles, bones, immune system, and nerves—changes during weightlessness and how the mind reacts to confinement and isolation can humanity ever hope to live in space and to journey to Mars.[11] One of the important challenges, everyone recognizes, is how to readjust to gravity when returning to Earth, for currently the astronauts undergo a lengthy period of recovery during which they must regain not only their strength but also their balance and coordination.[12]

Many of the physiological changes in astronauts actually resemble changes associated with aging. For instance, in addition to losing mass in the microgravity environment, bones and muscle do not appear to heal normally in space. Time spent in microgravity seems to result in a dissociation between physical and chronological age. By studying the changes in astronauts' bodies, therefore, ISS research might play a role in developing a model for the consequences of getting older. NASA-sponsored scientists are collaborating with the National Institutes of Health in an effort to explore the use of spaceflight as a model for the process of aging.[13]

Researchers have found that microgravity provides them with new tools to address two fundamental aspects of biotechnology: the growth of high-quality crystals for the study of proteins, and the growth of three-dimensional tissue samples in laboratory cultures. On Earth gravity distorts the shape of crystalline structures, while tissue cultures fail to take on their full three-dimensional structure. Research on the station may provide new data and techniques for answering the questions of the broader biotechnology community. NASA's bioreactor, developed to simulate low gravity, has proven dramatically successful as an advanced cell-culturing technology. This has led to an extensive collaboration between NASA, NIH, and numerous commercial partners. Work with bioreactors has already produced advanced cultures of lymph tissue for studying the infectivity of the human immunodeficiency virus (HIV), the one that causes AIDS. Other areas of success include cultures of cancer tumors and the growth of cartilage.

Biotechnology researchers use microgravity as well to produce for drug research protein crystals that are superior to to those grown on Earth. Using high-energy X-ray beams to study high-quality crystals, scientists are better able to discern how the proteins function. Work done by researchers and pharmaceutical companies on space-grown crystals has already increased our knowledge about AIDS, emphysema, influenza, and diabetes.

Commercial, academic, and government researchers have successfully used the low-gravity environment of space to understand and control gravity's influence in the production and processing of materials including metals, semiconductors, polymers, and glasses. For example, space research has produced cadmium zinc telluride crystals with fifty times lower levels of a key defect than the best commercially available crystals. These experiments may ultimately help researchers improve semiconductor fabrication on Earth.

Initial ISS research will be aimed at validating the research facility's performance and will provide baseline data. As more capabilities evolve on the ISS, scientists intend to develop a larger scientific research agenda. They envision cycling experiments through ISS over a four-month time frame, using the crew rotations and flights to and from ISS to plan and end experiments. Each four-month period will emphasize a theme that focuses on a primary scientific discipline. The first several increments will involve biomedical research, effects of radiation on humans, research of bone and muscle, plants in space, and "from molecules to matter," using space to probe the forces that structure Earth.[14] As recently as September 16, 2002, NASA took an important step forward in naming an ISS science officer to oversee all experiments on board the station. NASA administrator Sean O'Keefe named astronaut Peggy Whitson of the Expedition 5 crew as the first science officer. The science officer, O'Keefe said, would concentrate exclusively on enhancing the scientific return from the ISS.[15]

Planning for research on the ISS has led to a series of broad principles informing the larger scientific agenda:

1. Broad participation by the world's community of researchers from academia, government agencies, and industry
2. Leverage of resources and expertise through agreements with various government, university, and research organizations to broaden the effectiveness of ISS programs
3. Emphasis on investigator-initiated research portfolio (the research done in space is determined by demand—through investigator's competed research proposals)
4. Ensuring of peer review of supported science and technology research projects selected from open competition
5. Optimized productivity of the highly diverse range of scientific, technological, and commercial investigations through optimal ISS research implementation
6. Approximately 30 percent of space station resources devoted to commercial users
7. Evaluation of performance, ultimately in terms of long-term impacts on our society, development of products to better life on Earth, and advances that will enable human exploration of space[16]

At least for now the Destiny Laboratory Module will be the centerpiece of NASA's ISS scientific research. Destiny was launched into space aboard Shuttle mission STS-98 on February 7,

2001, and has a capacity of twenty-four racks occupying thirteen locations specially designed to support experiments. More than fifty experiments had been completed by the end of 2002, ranging from materials science to biology.[17]

To manage the scientific activity on the ISS, NASA has been considering the creation of a non-government organization (NGO). Because of the complexity of the effort, this may prove appropriate. The ISS will include more than twenty-five internal laboratory sites and twenty-five external platform sites to support U.S. research and development projects. The U.S. utilization program will also require extensive coordination of researchers from academia, industry, and government, as well as close liaison to the programs of our international partners in Canada, Europe, Japan, and Russia. A separate entity overseeing this research may alleviate some of the burden of management that would otherwise fall to NASA. A set of reports on the NGO option was completed in 2002, and a final decision on the matter should be forthcoming in 2003.[18]

The key to successful scientific investigations on the ISS is the Destiny Laboratory Module. Here is an overall shot of the newly attached Destiny, recorded in February 2001 with a 35mm camera during the early occupancy by astronauts and cosmonauts from the Expedition One and STS-98 crews. (NASA photo, no. STS098-355-0008)

Astronaut Susan J. Helms, Expedition Two flight engineer, works at the Human Research Facility's ultrasound flat screen display and keyboard module in the Destiny Laboratory. (NASA photo, no. ISS002-E-6699)

THE ISS AND THE CLASH OF CIVILIZATIONS

The space station may well provide an opportunity for the completion of a multitude of scientific experiments, but that might not be the most important reason for supporting it. At a fundamental level, the geopolitical system of the post–Cold War era requires its continuation, along with the establishment of other large-scale international programs that enhance engagement with other civilizations. Although difficult, it is necessary. One is reminded of the quote from Wernher von Braun, "We can lick gravity, but sometimes the paperwork is overwhelming."[19] Perhaps the hardest part of spaceflight is not the scientific and technological challenges of operating in an exceptionally foreign and hostile environment but the down-to-Earth environment of rough-and-tumble international and domestic politics.

But even so, cooperative space endeavors have been richly rewarding and overwhelmingly useful, from all manner of scientific, technical, social, and political perspectives. This is especially

Astronaut Daniel W. Bursch, Expedition Four flight engineer, performs a cardiopulmonary resuscitation experiment on a jerry-rigged "human chest" dummy in the Destiny Laboratory, March 11, 2002. (NASA photo, no. ISS004-E-8505)

true of the International Space Station. Virtually everyone would agree that astronauts standing on the Moon alongside the United States flag were just as important to the winning of the Cold War as were ballistic missiles and strategic weapons. Just as surely as the Apollo program helped to win the Cold War for the United States, the ISS serves a critical international role in the post–Cold War world.[20]

In the aftermath of international tensions, the International Space Station may prove just as important in the quest to maintain U.S. hegemony—political, technological, and economic—as Apollo was at the height of the Cold War. Since the collapse of the Soviet Union, a different set of priorities has replaced the powerful secular ideologies of democracy, communism, nationalism, fascism, and socialism that had dominated international politics since the Enlightenment. They are not so much new priorities as ancient traditions based on ethnic, religious, kinship, and tribal loyalties, and they reemerged full-blown in the 1990s as all the great ideologies except democracy collapsed worldwide—and even democracy was none too stable outside of the West.[21]

Harvard political scientist Samuel P. Huntington developed a powerful thesis to explain what has happened since the end of a bipolar world. The thrust of his argument rejects the notion that the world will inevitably succumb to Western values. On the contrary, Huntington contends that the West's influence is waning because of growing resistance to its values and the reassertion by non-Westerners of their own cultures. He argues that the world will see in the twenty-first century an increasing threat of violence arising from renewed conflicts between countries and cultures basing their identities on long-held traditions.

This argument moves past the notion of ethnicity to examine the growing influence of a handful of major cultures—Western, Orthodox, Latin American, Islamic, Japanese, Chinese, Hindu, and African—in current struggles across the globe. Thus Huntington successfully shifts the discussion of the post–Cold War world from ideology, ethnicity, politics, and economics to culture—especially its religious basis. Huntington rightly warns against facile generalizations about the world becoming one, so common in the early 1990s, and points out the resilience of civilizations to foreign secular influences.[22]

Huntington asserts that there are nine major civilizations in the post-1990 era. The dominant at present is the West, characterized by the United States, Canada, and the nations of Western Europe. Aside from this there are Latin American, African, Islamic, Sinic (Chinese), Hindu, Orthodox (Russian and other Slavic nations), Buddhist, and Japanese civilizations. Each has different traditions, priorities, and institutions. Each also misunderstands the other civilizations of the world. In the post–Cold War era, no matter how seemingly desperate confrontations within these civilizations may seem—such as the trials over northern Ireland—they have little potential for escalation beyond the civilization in which they occur. Confrontations between civilizations, however, have a large potential to escalate into large conflagrations, even world wars. Those capable of forming meaningful ties to others, creating alliances not just for defensive purposes but also as a means of broadening engagement, have the greatest possibility for thriving in this new international arena. The West, Huntington believes, should give up the idea of exporting its values and expand the possibility of its survival through stronger alliances with other civilizations.[23]

In the clash of civilizations of the twenty-first century, the International Space Station offers a test bed for civilizational alliances. At some level this has already begun. From the beginning, the West adopted the project and invited another great civilization in Japan to participate. In 1993 the "Orthodox civilization," using Huntington's terminology for Russia and other Slavic peoples, joined the program. Perhaps the difficulty of working with the Russians has been largely the result of these strikingly different civilizations. Brazil and other nations of the Latin American civi-

In the 1970s NASA's Ames Research Center undertook two space colony summer studies in which Gerard K. O'Neill participated. Communities housing about 10,000 people were designed. A number of artistic renderings were made, and shown here is a Bernal Sphere, with a living area that includes a human-powered airplane. (NASA photo, no. AC76-0628)

lization also want to join the program, as does India. China has made overtures about the desire to become a part of the ISS effort. Despite the very real challenges that would result from incorporating these new partners into the program, their inclusion would advance the cause of creating alliances with other civilizations, which could serve ultimately as a means of closing the gap between nations rather than widening it. At a fundamental level, the International Space Station would serve the larger objectives of American foreign policy better than many other initiatives with dimmer prospects for success.[24]

All the promise held out for the ISS in gaining scientific knowledge, technological development, and exploration of the solar system may well pale in comparison to the very real possibility of enhancing cross-civilizational relations through this one act of working together to tackle an enormous challenge. The same may be true of the very real costs involved, a small price to pay for better international relations. And spending a larger share of the public treasury for the ISS is imminently better than spending it for weapons of destruction. For all the difficulties involved, the knowledge gained in large-scale cooperative programs will serve the United States and the West well in the inter-civilizational rivalries of the twenty-first century.

Space Stations of an Imaginary Future

Space stations have long shouldered the hopes and dreams of generations of humans longing to explore the heavens. By the turn of this century those dreams may have been deferred but not abandoned, and ISS may foster the knowledge necessary to make them a reality, if not in the immediate future then in the long term. The central question is this: What will have to be learned in order for humans to make space their home? It is a hostile environment out there, and the work conducted on the ISS is central to answering this question. That answer will make the ISS in reality a base camp to the stars.

In the 1970s Gerard K. O'Neill explored the possibility of colonies scattered throughout the solar system. In this capacity he left an indelible mark on the development of the rationale for a space station. He argued that the potentiality for human colonies in free space seemed limitless, as he calculated the technical issues of energy, land area, size and shape, atmosphere, gravitation, and sunlight necessary to sustain a colony in an artificial living space. Rather than live on the outside of a planet, settlers could live on the inside of gigantic cylinders or spheres of roughly one-half to a few miles in each dimension. These would hold a breathable atmosphere, all the ingredients necessary for sustaining crops and life, and be rotating habitats to provide artificial gravity. Although the human race might eventually build millions of such colonies, each settlement would of necessity be an independent biosphere, with trees and lakes and blue skies spotted with clouds along each colony's inner rim where all oxygen, water, waste, and other materials could be recycled endlessly. Animals and plants endangered on Earth would thrive on these cosmic arks; insect pests would be left behind. Solar power, directed into each colony by huge mirrors, would provide a constant source of nonpolluting energy.[25]

This bold vision catapulted O'Neill into the spotlight of the space community and received a major boost from numerous science fiction and science fact writers, among them Arthur C. Clarke, who popularized the concept.[26] O'Neill's vision of a practical and profitable colony in space found an audience in many quarters of NASA, as it did in the larger pro-space movement. He received funding from NASA's Advanced Programs Office—but only $25,000—to develop his ideas more fully. Senior NASA officials such as administrator James C. Fletcher and Ames Research Center director Hans Mark personally encouraged his efforts.[27]

In the summer of 1975, NASA officials took O'Neill's ideas seriously enough to convene a study group of scientists, engineers, economists, and sociologists at the Ames Research Center, near San Francisco, to review the idea of space colonization, and followed it up with a

This is Gerard K. O'Neill's "Island One," with its agricultural areas and radiators for waste heat. All energy for habitat, agriculture, and industries was to be derived from solar power. This type of structure was named the Bernal Sphere in honor of the British scientist-author J. D. Bernal, who wrote of the humanization of space in the 1920s. In actual construction, one layer of protective shell would be installed first, and then the interior sphere would be completed and filled with atmosphere before architectural work began. (Courtesy of Space Studies Institute, Princeton, N.J.)

study the next summer. Surprisingly they found enough in the scheme to recommend it. Although budget estimates of $100 billion accompanied any colonization project, the authors of this study concluded: "In contrast to Apollo, it appears that space colonization may be a paying proposition." For them, it offered "a way out from the sense of closure and of limits which is now oppressive to many people on Earth." Most importantly, the study concluded: "The possibility of cooperation among nations, in an enterprise which can yield new wealth for all rather than a conflict over the remaining resources of the Earth, may be far more important in the long run than the immediate return of energy of the Earth. So, too, may be the

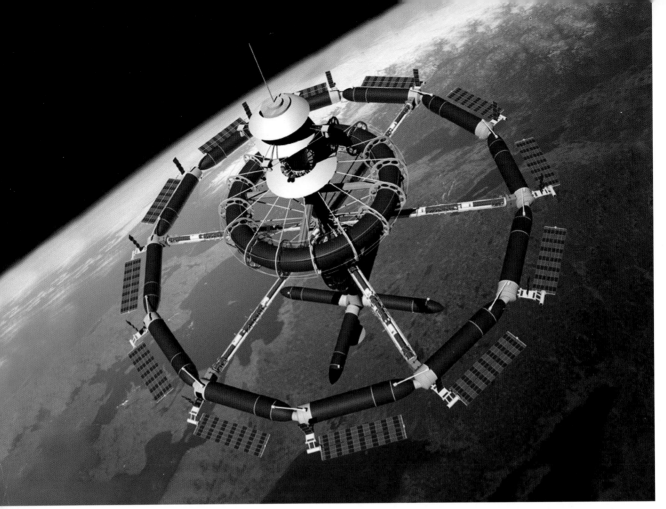

One of the most persistent objectives of spaceflight has been an eventual capability for tourism. Here is a recent artist's conception of a wheeled space station constructed from discarded 28-foot by 180-foot External Fuel Tanks (ETs) used by the Space Shuttle. Championed by the Space Island Group, this concept features a dozen or more ETs joined into a slowly rotating station, with interiors outfitted to resemble the decks of small cruise ships accommodating 500 guests and staff. Occupants would enjoy one-third gravity levels, while the center of the ring would be gravity-free. By 2012 the Space Island Group plans to place into operation one of these wheel-shaped stations in an equatorial about 400 nautical miles above the Earth. The use of expended fuel tanks was first suggested for NASA's Skylab program, and the idea has been revisited during the Shuttle era. (Courtesy of Space Island Group)

sense of hope and of new options and opportunities which space colonization can bring to a world which has lost its frontiers."[28]

O'Neill publicized these findings exhaustively, but with political will for an aggressive space effort at low tide in the latter 1970s, nothing came of it. Both O'Neill and his supporters criticized NASA for not turning the dream into reality.[29] This created a wedge between prospace advocates and the government agency that was charged with space exploration.

Following closely on the heels of the two space colonization summer studies, in 1979 O'Neill founded the Space Studies Institute at Princeton University, with the intention of organizing small groups into teams to develop the "tools of space exploration independently of governments and to prove that private groups could get things done enormously cheaper and quicker than government bureaucracies." He never strayed from his belief that the private sector would be the only way to open the space frontier through colonization. Indeed he was appointed to the 1985 National Commission on Space in no small part because of his expansive vision of space colonization and because of his increasingly dogged commitment to space entrepreneurship. Freeman Dyson wrote of O'Neill after his death in 1992:

> I was privileged to be a close friend of two great men, [Nobel laureate in physics] Richard Feynman and Gerard O'Neill. I was often struck by the deep similarity of their characters, in spite of many superficial differences. Both were indefatigable workers, taking infinite trouble to get the details right. Both were effective and enthusiastic teachers. Both were accomplished showmen, good at handling a crowd. Both had good rapport with ordinary people and abhorred pedants and snobs. Both were uncompromisingly honest. Both were outsiders in their own profession, unwilling to swim with the stream. Both stood up against the established wisdom and were proved right. Both fought a fatal illness for the last seven years of their lives. Both had spirits that grew stronger as their physical strength decayed.

Freeman Dyson credited O'Neill with rescuing space colonization from the backwater of crackpots and making it the centerpiece of a hopeful future for humanity.[30]

O'Neill's ideas are present in the space community right up the present. At their core is the space station as a first step toward human settlements in space. Ultimately O'Neill may have been right, but in the short run there are significant hurdles to be overcome.

A Beginning?

At the least, the ISS should open more activity in low Earth orbit than ever seen before. With the launch of the first ISS elements in 1998, John M. Logsdon concluded, "There is little doubt, then, that there will be an International Space Station, barring major catastrophes like another Shuttle accident or the rise to power of a Russian government opposed to cooperation with the West."[31] Logsdon noted that it was remarkable that the space station program had survived to that point, because of its weak support over the years both internationally and domestically. He also commented: "Even with all its difficulties and compromises, the space station partnership still stands as the most likely model for future human activities in space. The complex multilateral mechanisms for managing station operations and utilization will become a de facto world space agency for human space flight operations, and planning for future missions beyond Earth orbit are most likely to occur within the political framework of the station partnership."[32]

Like earlier spaceflight projects, especially Apollo, the most important legacy of ISS may be what it teaches humanity about managing complex multinational projects. The ability to build modules all over the world, to assemble them on orbit, and to have them work properly is both amazing and instructive for future generations. NASA engineers hope that the station, once functioning in space, may foster the building of free-flying laboratories operated by private companies. Daimler-Chrysler, for instance, has a contract to build one module. That first module could cost more than $1 billion, but what would the second, third, or fourth cost? Mass-producing modules originally built for the ISS could open new avenues for space operations. High-tech tenants of an orbital "research park" could grow up around the ISS "anchor" research facility to take advantage of the microgravity environment of space. Pharmaceutical research in this environment holds especially great promise, but structures and materials research is also enhanced in microgravity. The value of this capability has already been proven on a limited basis with the Space Shuttle, and the ISS holds the potential to "kick-start" commercial research.[33]

Although the majority of research modules might be uninhabited—after all, the delicate nature of the research could be jarred by human presence—they will require periodic attention from professional teams of researchers in low Earth orbit. Such a staff, employed by corporations rather than by NASA, may well become the first routine spacefarers, operating like scientists in Antarctica spending months on the continent and then returning home for extended stays. They will reside in habitat modules either connected to the ISS or in close proximity to it.

But if research teams can travel back and forth between a habitat module in the heart of this orbiting research park, why not dedicate a habitat module to those traveling for pleasure? One of the compelling visions of the space station as a base camp to the stars is its role as a resort for space tourists. Visions of anyone with significant resources being able to travel into orbit and enjoying a vacation on a space station or at a lunar resort abound in fiction. The classic representation is Stanley Kubrick's large wheeled space station shown in the film *2001: A Space Odyssey*, released in 1968. A massive structure that not only provided a base for research but also served as a way station for travelers on their way to the Moon and a resort for tourists offering amenities as comforting as those in a modern hotel.[34]

Numerous space station enthusiasts believe that the realization of the ISS offers a unique possibility for space tourism to become a reality, after more than forty years of control over access to space held by the governments of the United States and the Soviet Union/Russia. Indeed, that was the premise of MirCorp, an Amsterdam-based international company created in the mid-1990s to use the Russian Mir space station as a hotel for high-priced vacations once it had been abandoned by the Russian Space Agency. The company even planned to create a new reality television series modeled on NBC's *Survivor,* which producer Mark Burnett called "Destination Mir." The winner would become a tourist for a week on Mir. Although these efforts went nowhere and Mir was deorbited in 2001, the dream remains very much alive.[35]

Several futurists believe that by the year 2030 there will be space tourists taking their vacations, albeit exceptionally expensive ones, in low Earth orbit. Market studies suggest that there are already more than a thousand people per year willing to spend $1 million each for a weekend in space. Even at multimillion-dollar prices it could become a billion-dollar-per-year business, space economist Patrick Collins believes, and it will grow significantly in the future. If the cost of a space vacation dropped to about $25,000 per person, the number of people making the flight would rise to about 700,000 each year, he predicts. This represents a revenue stream of $17.5 billion per year.[36]

The industry is already beginning to see the first space tourists, as Dennis Tito pioneered the way by spending a week in 2001 on the ISS. In so doing, advocates of space tourism believe, he has challenged and overturned the dominant paradigm of human spaceflight: national control of who flies in space overseen with a heavy hand by NASA and the Russian Space Agency.

Dennis Tito's saga began in June 2000, when he signed a deal with MirCorp to fly aboard a Soyuz rocket to Mir. MirCorp acted as Tito's broker with the Russian space firm Energia, which owned both Mir and the rocket that would get Tito into space. Although MirCorp had grandiose plans for operating a space station supporting tourists and commercial activities, its leaders failed to obtain the venture capital necessary to make it a reality. Despite their efforts, they could not raise enough money to keep Mir in orbit, and the Russians announced in December 2000 that they would deorbit the space station.

This forced Tito to look elsewhere for a trip into space, and he negotiated with the Russians to fly to the ISS. Although the cash-starved Russian Space Agency was happy to make this deal, no one bothered to discuss it with any of the international partners building the ISS. A meltdown in public relations ensued, and NASA led the other partners in a rebellion that reached high into the political systems of the United States and Russia. NASA tried to persuade Tito to postpone his flight in February 2001, ostensibly to undergo two months of additional training before flying in October, but really to win time to convince the Russians not to allow Tito to fly. NASA and the other international partners argued that this slippage was paramount because of safety considerations on orbit.

Ever a cagey gamester, Tito saw the trap and refused. In March he forced a confrontation with NASA at the gates of the Johnson Space Center, where he planned to undergo training in preparation for an April 2001 flight. NASA lost that argument and was demonized by space enthusiasts for trying to block access to space for ordinary tourists. The Johnson Space Center acting director, Roy W. Estess, reflected a year later that he and his staff had not handled the Tito episode well and would have been better off to embrace the effort, while always ensuring the safety of the mission.[37]

With that one incident in Houston, Tito became a cause célèbre among space activists and NASA haters, who viewed him as the vanguard of a new age of space for everyone. Space psychologist Albert A. Harrison summarized the beliefs of many when he opined that "Tourism is one of the world's largest industries, and Russia's sale of a twenty-million-dollar space station trip to Dennis Tito represents but the first attempt to pry open the door for civilians in space. (Is there an irony here that the Russians are the entrepreneurs prying open the door for space tourism while the Americans try to preserve a government monopoly?)"[38] A SPACE.com web site visitors' poll taken in early May 2001—which did not represent a random sample by any means, but suggested where the space enthusiasts came down on the issue—showed that 75 percent of respondents supported Tito's flight, 24 percent believed he should not have flown, and 1 percent were undecided.[39]

Tito would not allow anything or anyone to stand in his way, and many space activists cheered as he thumbed his nose at "big, bad NASA" to take his week-long vacation on the ISS at the end of April 2001. In making his way over the objections of NASA, Tito may have paved the way for other millionaires to follow. South African millionaire Dennis Shuttleworth also flew aboard ISS in the fall of 2001, without the rancor of the Tito mission. Others will make the excursion in the future, either paying their own way or obtaining corporate sponsorships. *NSync singer Lance Bass made an abortive attempt in the fall of 2002, perhaps someday to succeed. But space policy analyst Dwayne A. Day does not believe this is the best way to open the space frontier. He wrote, "Now that Tito has flown, it will not be the Earth-shattering precedent that space enthusiasts hoped for . . . is it any easier for the average citizen to raise $20 million in cash and buy a seat on a Soyuz than it is to get a Ph.D. in engineering and join the astronaut corps? No. Far from opening a frontier, Tito's flight symbolizes just how far out of reach space remains for the common person."[40]

The flight of Dennis Tito offers an ambivalent precedent for the opening of space flight for the average person. Space tourism seems little closer today, even with the ISS, than it did in earlier eras. If there is a way to bring down the cost of access to space, then this dynamic may change. But until then, it does not much matter how many space stations are in orbit. Without a convenient, safe, reliable, and less costly means to reach them, little will change.[41] Once less expensive access to space is attained, an opening of the frontier might take place in much the same way as the American continental frontier emerged in the nineteenth century: through a linkage of courage and curiosity with capitalism. As this movement grows, the role of the government will become less dominant in space. NASA will continue research and development for space systems and carry out far-reaching science activities. But widespread human spaceflight could become the province of the commercial sector in the first half of the twenty-first century.

Only at that point might one conclude that the space station is truly serving as a base camp to the stars, as its earliest advocates wanted. Much will take place between the present and that seemingly far off future, not all of it positive and perhaps none of it linear in development. But this path into the future, although seemingly stalled, may already be in the process of becoming.

APPENDICES

Appendix 1

Manufacturers and Basic Characteristics of Skylab

Module	Manufacturer	Function
Apollo Command and Service Module	Rockwell International	Crew ascent and descent, attitude control
Docking Adapter	Martin Marietta	Docking interface, controls and displays, Earth observation, stowage
Solar Observatory (Apollo Telescope Mount)	MSFC	Solar observation, power source, attitude control
Airlock	McDonnell Douglas–Eastern	Power control and distribution, environmental control, data center, extravehicular activity hatch, caution and warning
Workshop	McDonnell Douglas–Western	Primary living and working area, laboratory, stowage, power source

Source: Leland F. Belew, *Skylab: Our First Space Station* (Washington, D.C.: NASA SP-400, 1977), 20.

Appendix 2

Dimensions of Skylab

Length	10.45 m (34.3 ft)	5.27 m (17.3 ft)	4.05 m (13.3 ft)	5.36 m (17.6 ft)	14.66 m (48.1 ft)
Diameter	3.96 m (13.0 ft)	3.04 m (10.0 ft)	3.35 m (11.0 ft)	3.04/1.67/6.70 m (10.0/5.5/22.0 ft)	6.70 m (22.0 ft)
Habitable working volume	5.95 m^3 (210 ft^3)	32.28 m^3 (1,140 ft^3)		17.66 m^3 (624 ft^3)	295.23 m^3 (10,426 ft^3)

Source: Leland F. Belew, *Skylab: Our First Space Station* (Washington, D.C.: NASA SP-400, 1977), 20.

Appendix 3
Skylab Extravehicular Activity

Mission	EVA Date	Flight Day	Duration (hr:min)	Crew	Primary Activity
Skylab 2	May 25, 1973	1	:40	Weitz	Stand-up EVA in unsuccessful attempt to free solar panel.
	June 7, 1973	13	3:25	Conrad, Kerwin	First successful repair EVA to free solar panel.
	June 19, 1973	25	1:36	Conrad, Weitz	Replacement of film canisters on Apollo Telescope Mount.
Skylab 3	Aug. 6, 1973	9	6:31	Garriott, Lousma	Deployment of sunshade to cool station.
	Aug. 24, 1973	27	4:31	Garriott, Lousma	Replacement of film canisters on Apollo Telescope Mount.
	Sept. 22, 1973	56	2:41	Bean, Garriott	Replacement of canisters on Apollo Telescope Mount.
Skylab 4	Nov. 22, 1973	6	6:33	E. Gibson, Pogue	Repair of antenna.
	Dec. 25, 1973	39	7:01	Carr, Pogue	Observation of Comet Kohoutek.
	Dec. 29, 1973	43	3:29	Carr, E. Gibson	Observation of Comet Kohoutek.
	Feb. 3, 1974	79	5:19	Carr, E. Gibson	Retrieval of film canisters and samples from station exterior.

Source: Aeronautics and Space Report of the President, 1999 Activities (Washington, D.C.: NASA Annual Report, October 2001), appendix C.

Appendix 4
Skylab Missions, 1973–74

Spacecraft	Launch Date	Crew	Flight Time (days:hrs:mins)	Highlights
Skylab 2	May 25, 1973	Charles Conrad Jr., Joseph P. Kerwin, Paul J. Weitz	28:0:50	Docked with Skylab 1 (launched uncrewed May 14) for 28 days. Repaired damaged station.
Skylab 3	July 28, 1973	Alan L. Bean, Jack R. Lousma, Owen K. Garriott	59:11:9	Docked with Skylab 1 for more than 59 days.
Skylab 4	Nov. 16, 1973	Gerald P. Carr, Edward G. Gibson, William R. Pogue	84:1:16	Docked with Skylab 1 in long-duration mission; last of the Skylab program.

Source: Aeronautics and Space Report of the President, 1999 Activities (Washington, D.C.: NASA Annual Report, October 2001), appendix C.

Appendix 5
International Crew Members Flown on the Salyut Space Stations

Crew	Affiliation/Country	Mission/Station	Launch Date
Vladimir Remek	Czechoslovakia	Soyuz 28/Salyut 6	Mar. 2, 1978
Miroslav Hermaszewski	Poland	Soyuz 30/Salyut 6	June 27, 1978
Sigmund Jŝhn	E. Germany	Soyuz 31/Salyut 6	Aug. 26, 1978
Georgiy Ivanov	Bulgaria	Soyuz 33	Apr. 10, 1979
Bertalan Farkas	Hungary	Soyuz 36/Salyut 6	May 25, 1980
Pham Tuan	Vietnam	Soyuz 37/Salyut 6	July 23, 1980
Arnaldo Tamayo Mendez	Cuba	Soyuz 38/Salyut 6	Sept. 18, 1980
Jugderdemidiyn Gurragcha	Mongolia	Soyuz 39/Salyut 6	Mar. 22, 1981
Dumitru Prunariu	Romania	Soyuz 40/Salyut 6	May 14, 1981
Jean-Loup Chrétien	France	Soyuz T6/Salyut 7	June 24, 1982
Jean-Loup Chrétien	France	Soyuz TM7/Mir	Nov 26, 1988
Rakesh Sharma	India	Soyuz T11/Salyut 7	Apr. 3, 1984

Source: Arnauld E. Nicogossian, Carolyn Leach Huntoon, and Sam L. Pool, eds., *Space Physiology and Medicine* (Philadelphia, Pa.: Lea and Febiger, 1994, 3d ed.), 24.

Appendix 6
Major Program Changes to U.S. Portion of Space Station Freedom

Calendar Year	Nature of Change	Reason
Fall 1985– May 1986	The original space station concept envisaged three elements: an occupied base for 8 crew members in a 28.5° orbit, an automated co-orbiting platform nearby, and an automated "polar platform" in orbit around Earth's poles. The original reference design for the occupied base was called the Power Tower, but a "dual-keel" approach was chosen instead as the baseline design in the fall of 1985; the details were approved by NASA in May 1986. Changes included: arrangement of truss structure and modules modified to place modules at center of gravity; solar dynamic power added to photovoltaic arrays; number of U.S. laboratory and habitation modules reduced from 4 to 2, with plans for 2 more provided by Europe and Japan (the new U.S. modules would be larger than the original design, however, so total habitable volume relatively unchanged); U.S. Flight Telerobotic Servicer added at congressional urging to supplement Canada's planned Mobile Servicing System.	Cost and user requirements. NASA stated that the dual-keel design would provide a better microgravity environment for scientists, more usable area for attached payloads, and better pointing accuracy. Cost estimate maintained at $8 billion (FY1984).
Late 1986	Dual-keel design reaffirmed, but emphasis on building single-keel first in recognition of reduced availability	January 1986 Space Shuttle Challenger tragedy and concern by astronauts at Johnson Space

Continued on next page

Calendar Year	Nature of Change	Reason
	of Shuttle flights and reduced amount of cargo that would be allowed aboard the Shuttle in the wake of the Challenger tragedy. Emphasis on early accommodation of experiments; fewer spacewalks; extended "safe haven" concept with the possibility for "lifeboats" for emergency return to Earth (not made a requirement at this time, reportedly for cost reasons); increased use of automation and robotics; "lead center" management approach replaced with dedicated space station program office in Reston, Va.	Center about the number of hours of spacewalks, or EVAs; quality and quantity of living space; standard of safety for safe havens (to which astronauts would retreat in emergencies such as depressurization or dangerous sunspot activity); lack of lifeboats for emergency return to Earth when the Space Shuttle was not docked with the station. Cost estimate unchanged.
1987	Program split into Phase 1 and Phase 2, with single keel of occupied base built in Phase 1 and second keel delayed until Phase 2; polar platform part of Phase 1; co-orbiting platform and solar dynamic power pushed into Phase 2.	Rising program costs and expected budget constraints. Cost estimate had risen to $14.5 billion (FY1984) for research and development. New design estimated to cost $12.2 billion (FY1984) for Phase 1 and $3.8 billion (FY1984) for Phase 2, saving money in the near term but costing more in the long term.
1989	Phase 2 indefinitely postponed; polar platform transferred from space station program to NASA's Office of Space Science and Applications (was for Earth observation studies). Only remaining element was single-keel occupied base, divided into an initial phase with reduced capabilities (e.g., crew reduced to 4 from 8; electrical power reduced to 37.5 kW from 75 kW; use of open-loop instead of closed-loop life support system) and an "assembly complete" phase when "full capabilities" would be restored. NASA asserted that the capabilities envisioned in the 1987 Phase 2 program (dual-keel, etc.) could still "evolve" sometime in the future to support expeditions to the Moon and Mars.	Cost growth and expected budget constraints. NASA termed this a "rephasing." Cost for Phase I estimated at $19 billion real year dollars, or $13 billion FY1984, for research and development; NASA estimated total program costs through assembly complete at $30 billion real year dollars.
1990–91	U.S. modules reduced in size (from 44 ft to 27 ft); "pre-integrated truss" chosen in effort to reduce EVA requirements; total length reduced (from 493 ft to 353 ft); Flight Telerobotic Servicer canceled; crew size formally reduced to 4; electrical power reduced (from 75 kW to 56 kW); lifeboat added to the station's design but not included in the cost estimate; assembly complete designation abandoned with concept that station will continually evolve in an undefined and unbudgeted "follow-on phase."	Beginning in 1990, concerns developed over rising program costs, weight, insufficient electrical power, and too many EVAs for maintenance. In December 1990, NASA estimated program costs through assembly complete at $38.3 billion real year dollars. Congress directed NASA to restructure the station. New plan was released in March 1991. NASA stated it would cost $30 billion real year dollars through 1999, though this was no longer the time when assembly would be completed (see column to the left). General Accounting Office estimated total program costs through 30 years of operation at $118 billion.

Source: Marcia S. Smith, "NASA's Space Station Program: Evolution and Current Status," House Science Committee testimony, April 4, 2001, NASA Historical Reference Collection, NASA History Office, Washington, D.C.

Appendix 7

NASA's Cost Estimates for U.S. Portion of the Space Station, 1984–2001

Date	Amount	Comments
1984	$8 billion	Initial estimate for Space Station Freedom (FY1984 dollars, R&D only, no Shuttle launches).
Apr. 1987	$16 billion	In restructuring of Freedom program, it was split into two phases: $12.2 billion for Phase 1; $3.8 billion for Phase 2 (FY1984 dollars, R&D only, no Shuttle launches).
Apr. 1989	$30 billion	Space Station Freedom Phase 1 (real year dollars [RYD], through assembly complete, including Shuttle launches during assembly and other costs—such as the Flight Telerobotic Servicer and ground facilities). Phase 2 "indefinitely postponed," so not included in this or subsequent cost estimates.
Early 1990	$37 billion	Space Station Freedom (RYD, through assembly complete, including Shuttle launches during assembly and other costs).
Dec. 1990	$38.3 billion	Space Station Freedom (RYD, through assembly complete, including Shuttle launches during assembly and other costs).
Mar. 1991	$30 billion	Space Station Freedom (RYD, through permanent human capability, including Shuttle launches during assembly and other costs).
Nov. 1993	$17.4 billion	Following termination of Freedom and initiation of International Space Station (RYD, development costs through assembly complete, no Shuttle launches, includes costs for science experiments).
Mar. 1998	$21.3 billion	RYD, development costs through assembly complete, no Shuttle launches, includes costs for science experiments.
Apr. 1998	$24.7 billion	Not a NASA estimate. Independent "Cost Assessment and Validation Team" headed by Jay Chabrow estimated cost through assembly complete.
June 1998	$22.7 billion	RYD, development costs through assembly complete, no Shuttle launches, includes costs for science experiments. NASA did not accept the Chabrow figure but agreed the program would cost $1.4 billion more.
Feb. 1999	$23.4–26 billion	RYD, development costs through assembly complete, no Shuttle launches, includes costs for science experiments.
Feb. 2000	$24.1–26.4 billion	RYD, development costs through assembly complete, no Shuttle launches, includes costs for science experiments.
Mar. 2001 (under discussion)	$22–23 billion	Assumes termination of construction after completion of U.S. Core and attachment of European and Japanese lab modules (RYD, development costs through completion of the U.S. Core; no Shuttle launches; includes costs for science experiments, reduced 40 percent from previous estimates).

Sources: Prepared by Congressional Research Service using cost data from NASA. Cost estimates reflect NASA's budget books (e.g., p. SI-6 of FY2001 budget book). Estimates in real year dollars (RYD) reflect current and prior year spending, unadjusted for inflation, plus future year spending that includes a factor accounting for expected inflation. Marcia S. Smith, "NASA's Space Station Program: Evolution and Current Status," House Science Committee testimony, April 4, 2001, NASA Historical Reference Collection, NASA History Office, Washington, D.C.

Appendix 8
Shuttle-Mir Missions, 1995–98

Spacecraft	Launch Date	Crew	Flight Time (days:hrs:mins)	Highlights
STS-63 Discovery	Feb. 3, 1995	James D. Wetherbee, Eileen M. Collins, Bernard A. Harris Jr., C. Michael Foale, Janice E. Voss, Vladimir G. Titov (Russia)	8:6:28	Sixty-seventh STS flight. Primary objective: first close encounter in nearly twenty years between American and Russian spacecraft as a prelude to establishment of International Space Station. Shuttle flew close by Mir. Main payloads: Space-hab 3 experiments and Shuttle Pointed Autonomous Research Tool for Astronomy (SPARTAN) 204, Solid Surface Combustion Experiment (SSCE), and Air Force Maui Optical Site (AMOS) Calibration Test. Also launched very small Orbital Debris Radar Calibration Spheres (ODERACS).
Soyuz TM-21	Mar. 14, 1995	Vladimir Dezhurov, Gennadi Strekalov, Norman Thagard	Crew flew for different periods of time.	Thagard was the first American astronaut to fly on a Russian rocket and to stay on the Mir space station. Soyuz TM-20 returned to Earth on Mar. 22, 1995, with Valeriy Polyakov, Alexsandr Viktorenko, and Telena Kondakova. Polyakov set world record by remaining in space for 438 days.
STS-71 Atlantis	June 27, 1995	Robert Gibson, Charlie Precourt, Ellen Baker, Gregory Harbaugh, Bonnie Dunbar	9:19:22	Sixty-ninth STS flight and one hundredth U.S. human spaceflight. Docked with Mir space station. Brought up Mir 19 crew (Anatoly Y. Solovyev and Nikolai M. Budarin). Returned to Earth with Mir 18 crew (Vladimir N. Dezhurov, Gennady M. Strekalov, and Norman Thagard). Thagard set an American record by remaining in space for 115 days.
STS-74 Atlantis	Nov. 12, 1995	Kenneth D. Cameron, James D. Halsell Jr., Chris A. Hadfield, Jerry L. Ross, William S. McArthur Jr.	8:4:31	Seventy-third STS flight. Docked with Mir space station as part of International Space Station Phase 1 efforts. Brought supplies and equipment, as well as solar arrays and experiments.

Spacecraft	Launch Date	Crew	Flight Time (days:hrs:mins)	Highlights
STS-76 Atlantis	Mar. 22, 1996	Kevin P. Chilton, Richard A. Searfoss, Linda M. Godwin, Michael R. Clifford, Ronald M. Sega, Shannon W. Lucid	9:5:16	Seventy-sixth STS flight. Docked with Mir and left astronaut Shannon Lucid aboard. Brought supplies and equipment, including Mir Environmental Effects Payload. Also carried the Spacelab module.
STS-79 Atlantis	Sept. 16, 1996	William F. Readdy, Terrence W. Wilcutt, Jerome Apt, Thomas D. Akers, Carl E. Walz, John E. Blaha, Shannon W. Lucid	10:3:19	Seventy-ninth STS flight. Docked with Mir. Picked up astronaut Shannon Lucid, who set another American record with 188 days in orbit, and dropped off astronaut John Blaha.
STS-81 Atlantis	Jan. 12, 1997	Michael A. Baker, Brent W. Jett Jr., John M. Grunsfeld, Marsha S. Ivins, Peter J. K. Wisoff, Jerry M. Linenger	10:2:17	Eighty-first STS flight. Fifth Mir docking. Picked up John Blaha and left Jerry Linenger on Mir.
STS-84 Atlantis	May 15, 1997	Charles J. Precourt, Eileen M. Collins, C. Michael Foale, Carlos I. Noriega, Edward T. Lu, Jean-François Clervoy, Telena V. Kondakova	9:6:42	Eighty-fourth STS mission. Nineteenth flight OV-104 Atlantis. Sixth Mir docking mission, exchanging crew members. Picked up Jerry Linenger and left Michael Foale.
STS-86 Atlantis	Sept. 25, 1997	James D. Wetherbee, Michael J. Bloomfield, Vladimar G. Titov, Scott E. Parazynski, Jean-Loup Chrétien, Wendy B. Lawrence, David A. Wolf	11:14:27	Eighty-seventh STS mission. Seventh Mir docking mission, exchanging crew members. Picked up Michael Foale and left David Wolf.
STS-89 Endeavour	Jan. 22, 1998	Terrence W. Wilcutt, Joe F. Edwards Jr., James F. Reilly II, Michael P. Anderson, Bonnie J. Dunbar, Salizhan S. Sharipov, Andrew S. Thomas, David A. Wolf	8:19:47	Eighth Shuttle docking mission to Mir. Andrew Thomas replaced David Wolf on Mir. Shuttle payloads included Spacehab double module of science experiments.
STS-91 Discovery	June 2, 1998	Charles J. Precourt, Dominic L. Pudwill Gorie, Franklin R. Chang-Diaz, Wendy B. Lawrence, Janet Lynn Kavandi, Valery V. Ryumin, Andrew S. Thomas	9:19:48	Last of nine docking missions with Mir. Brought home Andrew Thomas. Payloads included the Department of Energy's Alpha Magnetic Spectrometer to study high-energy particles from deep space, four Get Away Specials, and two Space Experiment Modules.

Source: *Aeronautics and Space Report of the President, 1999 Activities* (Washington, D.C.: NASA Annual Report, October 2001), appendix C.

Appendix 9

NASA/Bush Administration Cost Estimates on International Space Station ($millions)

Item	FY2002	FY2003	FY2004	FY2005	FY2006	Total
FY2002 pres. request[a]	2,087.3	1,817.5	1,509.1	1,394.3	1,389.0	8,197.3
OMB/NASA adjustments	27.5	30.1	33.4	31.1	28.4	150.5
Shuttle X-38 flight transfer	8.5	11.1	6.4	4.1	1.4	31.5
HEDS tech transfer	19.0	19.0	27.0	27.0	27.0	119.0
NASA/OMB agreement	2,114.9	1,847.6	1,542.5	1,425.4	1,417.4	8,347.8
Remaining shortfall			−141.3	−194.6	−148.3	−484.2

Source: "House Science Committee Hearing Charter: The Space Station Task Force Report," Committee on Science, U.S. House of Representatives, November 7, 2001, NASA Historical Reference Collection.

[a] FY02 request deletes funding for CRV, Hab. Module, and U.S. Propulsion Module (limiting to a three-person crew).

Appendix 10

International Space Station Independent Management and Cost Evaluation Task Force Cost Estimates, November 1, 2001

Item	Budget	Shortfall/Offsets	Reserve
NASA/OMB agreement (FY02–06)	$8.3B	($484M)	$750M
Probable increases costed by task force		($366M)	
Offsets identified by task force		$110M	
Content changes and rates		$330M	
Operations and sustaining engineering		—	
Additional resources based on recommendations		$668M	
Reduced Shuttle flight rate		$350-	
Potential savings from institutional changes		$450M	
Task force recommended solution	$8.3B	$600M	$750M

Source: IMCE Task Force Report, November 1, 2001.

Note: Did not include costs for contractor rate changes, costs stemming from changes in the baseline, and research changes. The task force identified, but did not provide, an estimated cost for these potential risks.

NOTES

1. ORIGINS

1. There are several fine overviews of the early history of spaceflight: William E. Burrows, *This New Ocean: The Story of the First Space Age* (New York: Random House, 1998); Tom D. Crouch, *Aiming for the Stars: The Dreamers and Doers of the Space Age* (Washington, D.C.: Smithsonian Institution Press, 1999); T. A. Heppenheimer, *Countdown: A History of Space Flight* (New York: John Wiley and Sons, 1997); John M. Logsdon, gen. ed. *Exploring the Unknown: Selected Documents in the History of the U.S. Civil Space Program,* 5 vols. (Washington, D.C.: NASA SP-4407, 1995–2001); Frederick I. Ordway III, *Visions of Spaceflight: Images from the Ordway Collection* (New York: Four Walls Eight Windows, 2001); Frederick I. Ordway III and Randy Liebermann, eds., *Blueprint for Space: Science Fiction to Science Fact* (Washington, D.C.: Smithsonian Institution Press, 1992). A previous publication of my own on the subject is *Frontiers of Space Exploration* (Westport, Conn.: Greenwood Press, 1998).

2. This is the argument of Howard E. McCurdy, *Space and the American Imagination* (Washington, D.C.: Smithsonian Institution Press, 1997).

3. George H. Gallup, *The Gallup Poll: Public Opinion, 1935–1971* (New York: Random House, 1972), vol. 1, 875, 1152.

4. Edward E. Hale, "The Brick Moon," *Atlantic Monthly* 24 (October 1869): 451–60; (November 1869): 603–11; (December 1869): 679–88.

5. This had been an enormously difficult problem, and nu-

merous schemes had been devised to provide navigational aids, some inane, others just impractical. The obvious value of a fixed point in the sky that could be used to plot courses was seen by all. An entertaining discussion of this is available in Dava Sobel, *Longitude: The True Story of a Lone Genius Who Solved the Greatest Scientific Problem of His Time* (New York: Walker and Co., 1995).

6. Roger D. Launius and Howard E. McCurdy, *Imagining Space: Achievements, Predictions, Possibilities, 1950–2050* (San Francisco, Calif.: Chronicle Books, 2001), 83–84.

7. Richard J. H. Barnes and Roy Gibson, "An International Look at Global Navigation Satellite Systems," *Space Policy* 14 (August 1998): 189–92; Irene A. Miller, "GPS and Beyond: How the Aviation Industry Is Advancing the Usefulness of GPS," *Quest* 7 (winter 1998): 41–45.

8. See George W. Morgenthaler and M. Hollstein, eds., *Space Shuttle and Spacelab Utilization: Near-Term and Long-Term Benefits for Mankind* (San Diego, Calif.: Univelt, 1978); *Science in Orbit: The Shuttle and Spacelab Experience, 1981–1986* (Washington, D.C.: NASA, 1988).

9. On this technology and its use, see Heather E. Hudson, *Communication Satellites: Their Development and Impact* (New York: Free Press, 1990); Andrew J. Butrica, ed., *Beyond the Ionosphere: Fifty Years of Satellite Communication* (Washington, D.C.: NASA SP-4217, 1997).

10. Willy Ley, *Rockets, Missiles, and Space Travel* (New York: Viking Press, 1961), 322; Willy Ley, *Rockets and Space Travel: The*

Future of Flight beyond the Stratosphere (New York: Viking Press, 1947), 284.

11. For information on Tsiolkovskiy, see Launius, *Frontiers of Space Exploration,* 91–92; Adam Starchild, ed., *The Science Fiction of Konstantin Tsiolkovskiy* (Seattle: University Press of the Pacific, 1979); Konstantin E. Tsiolkovskiy, *Aerodynamics* (Washington, D.C.: NASA, TT F-236, 1965); Konstantin E. Tsiolkovskiy, *Works on Rocket Technology* (Washington, D.C.: NASA, TT F-243, 1965); Arkady Kosmodemyansky, *Konstantin Tsiolkovskiy* (Moscow: Nauka, 1985).

12. A. A. Blagonravov, editor in chief, *Collected Works of K. E. Tsiolkovskiy,* vol. 2, *Reactive Flying Machines,* translation of "K. E. Tsiolkovskiy. Sobraniye Sochineniy. Tom II. Reaktivnyye Letatel'nyye Apparaty" (Moscow: Izdatel'stvo Akademii Nauk, 1954; NASA TT F-237, 1965).

13. See A. A. Kosmodemyansky, *K. E. Tsiolkovskiy: His Life and Work* (Moscow: Nauka, 1960), 153; Anatoly Zak, "The Cradle of the Russian Rocket Science," *Journal of the British Interplanetary Society* 49 (July 1996): 242–48.

14. N. A. Rynin, *Interplanetary Flight and Communication,* vol. 3, no. 7, K. E. Tsiolkovskii, Life, Writings, and Rockets (Leningrad, 1931; NASA TT F-646, 1971), 24–25; I. A. Kol'chenko and I. V. Strazheva, "The Ideas of K.E. Tsiolkovskiy on Orbital Space Stations," in R. Cargill Hall, ed., *Essays on the History of Rocketry and Astronautics: Proceedings of the Third through the Sixth History Symposia of the International Academy of Astronautics,* vol. 1 (Washington, D.C.: NASA Conference Publication 2014, 1977), 171.

15. Tsiolkovskiy, *Reactive Flying Machines.*

16. K. E. Tsiolkovski, *Exploring Universal Expanses with Jet Instruments* (Moscow: Nauka, 1934).

17. Peter A. Gorin, "Rising from the Cradle: Soviet Perceptions of Space Flight before Sputnik," in Roger D. Launius, John M. Logsdon, and Robert W. Smith, eds., *Reconsidering Sputnik: Forty Years since the Soviet Satellite* (Amsterdam: Harwood Academic Publishers, 2000), 11–42.

18. Hermann Oberth, "From My Life," *Astronautics* (June 1959): 39.

19. Hermann Oberth, *Ways to Spaceflight* (Washington, D.C.: NASA, TT F-622, 1972); Hermann Oberth, *Rockets into Planetary Space* (Washington, D.C.: NASA, TT F-9227, 1972); H. B. Walters, *Hermann Oberth: Father of Space Travel* (New York: Macmillan, 1962).

20. See Adam Gruen, "The Port Unknown: A History of the Space Station Freedom Program," unpublished manuscript (NASA Historical Reference Collection, NASA History Office, Washington, D.C., 1992), 1–2.

21. The standard work on the rocket societies is Frank H. Winter, *Prelude to the Space Age: The Rocket Societies, 1924–1940* (Washington, D.C.: Smithsonian Institution Press, 1983).

22. Hermann Oberth to Adam L. Gruen, May 14, 1985, NASA Historical Reference Collection.

23. This German effort to build the V-2 has been authoritatively discussed in Michael S. Neufeld, *The Rocket and the Reich: Peenëmunde and the Coming of the Ballistic Missle Era* (New York: Free Press, 1995).

24. Launius, *Frontiers of Space Exploration,* 87–89.

25. Hermann Noordung, *The Problem of Space Travel: The Rocket Motor,* eds. Ernst Stuhlinger and J. D. Hunley, with Jennifer Garland (Washington, D.C.: NASA SP-4026, 1995).

26. Frank H. Winter, "Observatories in Space, 1920s Style," *Griffith Observer* 46 (June 1982): 3–5; Fritz Sykora, "Pioniere der Raketentechnik aus Österreich," *Blätter für Technikgeschichte* 22 (1960): 189–92, 196–99; Ron Miller, "Herman Potočnik: *Alias* Hermann Noordung," *Journal of the British Interplanetary Society* 45 (1992): 295–96; Harry O. Ruppe, "Noordung: Der Mann und sein Werk," *Astronautik* 13 (1976): 81–83; Herbert J. Pichler, "Hermann Potočnik-Noordung, 22 Dez. 1892–27 Aug. 1929," typescript from a folder on Potočnik in the National Air and Space Museum's archive. Further information about early space stations as embarkation points appears in Barton C. Hacker, "The Idea of Rendezvous: From Space Station to Orbital Operations in Space-Travel Thought," *Technology and Culture* 15 (1974): 380–84. On Pirquet's series of articles, see also *Die Rakete: Offizielles Organ des Vereins für Raumschiffahrt E. V. in Deutschland* 2 (1928): esp. 118, 137–40, 184, and 189.

27. Frank H. Winter, *Rockets into Space* (Cambridge: Harvard University Press, 1990), 25–26.

28. Noordung, *Problem of Space Travel,* 101–2.

29. Ibid., 102–12.

30. Ibid., 112–13.

31. Ibid., 113.

32. Ley, *Rockets, Missiles, and Space Travel,* 369.

33. *Science Wonder Stories* 1 (July–September 1929): 170–80, 264–72, and 361–68. See also Tom D. Crouch, "'To Fly to the World in the Moon': Cosmic Voyaging in Fact and Fiction from Lucian to Sputnik," in Eugene M. Emme, ed., *Science Fiction and Space Futures, Past and Present,* AAS History Series, vol. 5 (San Diego, Calif.: Univelt, 1982), 19–22; Sam Moscowitz, "The Growth of Science Fiction from 1900 to the Early 1950s," in Ordway and Liebermann, eds., *Blueprint for Space,* 69–82.

34. J. D. Hunley, "Indroduction," in Noordung, *Problem of Space Travel.*

35. Winter, *Prelude to the Space Age,* 114; Gruen, "The Port Unknown," 13 n. 7; Wernher von Braun, "Crossing the Last Frontier," *Collier's* (March 22, 1952): 23, 29, 72–73; Wernher von Braun with Cornelius Ryan, "Can We Get to Mars?" *Collier's* (April 30, 1954): 22–28; Randy Liebermann, "The *Collier's* and Disney Series," in Ordway and Liebermann, eds., *Blueprint for Space,* 135–44.

36. Hacker, "Idea of Rendezvous," 384–85; Frederick I. Ordway III, "The History, Evolution, and Benefits of the Space Station Concept (in the United States and Western Europe),"

paper presented at the Thirteenth Congress of the History of Science, Section 12, Moscow, August 1971, p. 6, later published in the *Actes du XIIIe Congrès International d'Histoire des Sciences* 12 (1974): 92–132; Harry E. Ross and Ralph A. Smith, "Orbital Bases," *Journal of the British Interplanetary Society* 8 (January 1949): 1–19.

37. Hunley, "Introduction," in Noordung, *Problem of Space Travel.*

38. See Sylvia D. Fries, "2001 to 1994: Political Environment and the Design of NASA's Space Station System," in Marcel C. LaFollette and Jeffrey K. Stine, eds., *Technology and Choice: Readings from Technology and Culture* (Chicago: University of Chicago Press, 1991), 233–58; Alex Roland, "The Evolution of Civil Space Station Concepts in the United States," May 1983, Alex Roland biographical file, NASA Historical Reference Collection.

39. Ley, *Rockets, Missiles, and Space Travel,* 227.

2. THE SPACE STATION AND THE VON BRAUN PARADIGM

1. This story has been engagingly told in a masterful recollection, Homer H. Hickam Jr., *The Rocket Boys: A Memoir* (New York: Delacorte Press, 1998), and made into an exceptional feature film, *October Sky* (Universal Pictures, 1999).

2. See Dwayne A. Day, "The Von Braun Paradigm," *Space Times: Magazine of the American Astronautical Society* 33 (November–December 1994): 12–15; Dwayne A. Day, "Paradigm Lost," *Space Policy* 11 (August 1995): 153–59.

3. On Wernher von Braun, see Ernst Stuhlinger and Frederick I. Ordway III, *Wernher von Braun, Crusader for Space: A Biographical Memoir* (Malabar, Fla.: Krieger Publishing, 1994); Michael J. Neufeld, *The Rocket and the Reich: Peenemünde and the Coming of the Ballistic Missile Era* (New York: Free Press, 1995).

4. On these articles, see Randy Liebermann, "The *Collier's* and Disney Series," in Ordway and Liebermann, eds., *Blueprint for Space,* 135–44; "Man Will Conquer Space Soon" series, *Collier's* (March 22, 1952): 23–76ff; Wernher von Braun with Cornelius Ryan, "Can We Get to Mars?" *Collier's* (April 30, 1954): 22–28.

5. Von Braun, "Crossing the Last Frontier," 24–29, 72–74.

6. Liebermann, "The *Collier's* and Disney Series," in Ordway and Liebermann, eds., *Blueprint for Space,* 141; Ron Miller, "Days of Future Past," *Omni* (October 1986): 76–81.

7. Liebermann, "The *Collier's* and Disney Series," in Ordway and Liebermann, eds., *Blueprint for Space,* 144–46; David R. Smith, "They're Following Our Script: Walt Disney's Trip to Tomorrowland," *Future* (May 1978): 59–60; Mike Wright, "The Disney–Von Braun Collaboration and Its Influence on Space Exploration," paper presented at conference "Inner Space, Outer Space: Humanities, Technology, and the Postmodern World," February 12–14, 1993, copy in possession of author; Willy Ley, *Rockets, Missiles, and Space Travel,* 331.

8. *TV Guide,* March 5, 1955, 9.

9. See the sustained discussion of this subject in Francis Paul Prucha, *Broadax and Bayonet: The Role of the United States Army in the Development of the Northwest, 1815–1860* (Madison: State Historical Society of Wisconsin, 1953), 18–29, 110–22, 157–58, 188–89, 219–20; Arthur K. Moore, *The Frontier Mind: A Cultural Analysis of the Kentucky Frontiersman* (Lexington: University of Kentucky Press, 1957), 64–66; Jackson W. Moore, *Bent's Old Fort: An Archeological Study* (Denver: State Historical Society of Colorado, 1973); Robert B. Roberts, *Encyclopedia of Historic Forts: The Military, Pioneer, and Trading Posts of the United States* (New York: Macmillan, 1988).

10. McCurdy, *Space and the American Imagination,* 163–64.

11. Ibid., 164–65.

12. See Roger D. Launius, "NASA Retrospect and Prospect: Space Policy in the 1950s and 1990s," in Chris Hables Gray, ed., *Technohistory: Using the History of American Technology in Interdisciplinary Research* (Malabar, Fla.: Krieger Publishing, 1996), 215–32. See also Roland Huntford, *The Last Place on Earth* (New York: Atheneum, 1985); and McCurdy, *Space and the American Imagination,* 165–66.

13. Von Braun, "Crossing the Last Frontier," 28–29; Wernher von Braun, "Man on the Moon: The Journey," *Collier's* (October 18, 1952): 52; Robert R. Gilruth, "Manned Space Stations," *Spaceflight* (August 1969): 258; S. Fred Singer, *Manned Laboratories in Space* (New York: Springer-Verlag, 1969).

14. Wernher von Braun to Chesley Bonestell, December 26, 1951, Container 42, Von Braun Papers, Library of Congress, Washington, D.C.

15. Von Braun, "Crossing the Last Frontier," 72; Ordway, *Visions of Spaceflight,* 131–33. On Bonestell, see Ron Miller and Frederick C. Durant III with Melvin H. Schuetz, *The Art of Chesley Bonestell* (London: Paper Tiger, 2001).

16. "What Are We Waiting For?" *Collier's* (March 22, 1952); Oscar Schachter, "Who Owns the Universe?" *Collier's* (March 22, 1952): 36, 70–71.

17. Wernher von Braun to Aristid V. Grosse, 21 June 1952, A. V. Grosse File, 1951–1957, Container 42, Von Braun Papers, Library of Congress, Washington, D.C.; Wernher von Braun, "The Early Steps in the Realization of the Space Station," 1952, Guided Missile Development Group, Redstone Arsenal, Wernher von Braun Collection, Redstone Arsenal Army Period, Marshall Space Flight Center Archives, Huntsville, Ala.

18. H. H. Koelle, E. E. Engler, and J. W. Massey, "Design Criteria and Their Application to Economical Manned Satellites," in Institute of the Aeronautic Sciences (IAS)/RAND/NASA, *Proceedings of the Manned Space Stations Symposium* (Los Angeles: RAND, 1960), 6, 24–32.

19. Koelle, Engler, and Massey, "Design Criteria and Their

Application to Economical Manned Satellites," 25–33, 35. See also Jay Holmes, "Space Effort Hurt by Slow Decisions," *Missiles and Rockets* (May 1, 1961): 16, and H. H. Koelle Biographical Information File, NASA Historical Reference Collection.

20. Darrell C. Romick, "Preliminary Engineering Study of a Satellite Station Concept Affording Immediate Service with Simultaneous Steady Evolution and Growth," presented to the twenty-fifth anniversary meeting of the American Rocket Society, November 14–18, 1955, Chicago, available in NASA Langley Research Center Technical Library, Hampton, Va.

21. As examples of the misunderstanding of Romick's work, see "Goodyear Proposes Smaller Space Station," *Aviation Week and Space Technology* (October 14, 1957): 115; *Popular Science* (March 1958): 98; Jerry Grey, *Beachheads in Space: A Blueprint for the Future* (New York: Macmillan, 1983), 10.

22. Shirley Thomas, *Men of Space,* vol. 1 (Philadelphia: Chilton Co., 1960), 16.

23. S. Everett Gleason, "Discussion at the 329th Meeting of the National Security Council, Wednesday, July 3, 1957," July 5, 1957, p. 2, NSC Records, DDE Presidential Papers, Dwight D. Eisenhower Library, Abilene, Kans.

24. Krafft Ehricke, "Analysis of Orbital Systems," presented to the Fifth International Astronautics Federation Congress, 1954, Innsbruck, Austria, pp. 18–58, Langley Research Center Technical Library.

25. Ehricke, "Analysis of Orbital Systems," 18–25, 31, 36, 56.

26. Giovanni Caprara, *Living in Space: From Science Fiction to the International Space Station* (Willowdale, Ontario, Canada: Firefly Books, 2000), 26–27.

27. W. David Compton and Charles D. Benson, *Living and Working in Space: A History of Skylab* (Washington, D.C.: NASA SP-4208, 1983), 22–36.

28. Heinz Herman Koelle et al., "Project Horizon, Phase I Report: A U.S. Army Study for the Establishment of a Lunar Military Outpost," 3 vols. (Huntsville, Ala.: U.S. Ordnance Missile Command, June 8, 1959).

29. Wernher von Braun to Krafft Ehricke, 6 September 1957, A. V. Grosse File, 1951–1957, Container 42, Von Braun Papers, Library of Congress, Washington, D.C.

30. This is a point that I believe is obvious but not universally accepted. The classic work on the subject is John M. Logsdon, *The Decision to Go to the Moon: Project Apollo and the National Interest* (Cambridge: MIT Press, 1970).

31. C. E. Brown, William J. O'Sullivan Jr., and C. H. Zimmerman, "A Study of the Problems Related to High-Speed, High Altitude Flight," June 25, 1953, copy in Langley Research Center Technical Library under code CN-141, 504. On March 7, 1958, nearly seven months before NACA became NASA, Langley management formed a "manned satellite research program" that was meant to be "a point of departure in crystallizing an extended research program that Langley should follow in implementing manned-satellite flight." See F. L. Thompson to the Research Di-

vision Chiefs, "Formulation of a manned-satellite research program," March 7, 1958, copy in Floyd L. Thompson collection, Langley Historical Archives, NASA Langley Research Center, Hampton, Va.

32. Office of Program Planning and Evaluation, "The Long Range Plan of the National Aeronautics and Space Administration," December 16, 1959, NASA Historical Reference Collection.

33. J. W. Crowley, director of Aeronautical and Space Research, to Langley, "Research Steering Committee on Manned Space Flight," April 1, 1959; H. J. E. Reid to NASA, "Research Steering Committee on Manned Space Flight—Nomination of Langley representative," April 9, 1959; J. W. Crowley to Laurence K. Loftin Jr., May 1, 1959; J. W. Crowley to Langley, "Research Steering Committee on Manned Space Flight," May 1, 1959, Floyd L. Thompson collection, Langley Historical Archives, NASA Langley Research Center; "Minutes of the Research Steering Committee on Manned Space Flight, NASA Headquarters Office, Washington, D.C.," May 25–26, 1959, NASA Historical Reference Collection.

34. W. Ray Hook, "Space Stations—Historical Review and Current Plans," unpublished paper presented at the winter annual meeting of the American Society of Mechanical Engineers, Phoenix, Ariz., November 14–19, 1982, 3, copy in Langley Research Center Technical Library.

35. Rene A. Berglund, "Space Station Research Configurations," NASA Technical Note D-1504, August 1962, 3–6.

36. On Project Echo, see Donald C. Elder, *Out from Behind the Eight-Ball: A History of Project Echo,* AAS History Series, vol. 16 (San Diego, Calif.: Univelt, 1995).

37. See Loftin's testimony on inflatable structures in space to the *Hearing before the Committee on Science and Astronautics, U.S. House of Representatives, 87th Congress, First Session, May 19, 1961* (Washington, D.C.: Government Printing Office, 1961), 6.

38. Emanuel Schnitzer, "Erectable Torus Manned Space Laboratory," May 16, 1960; Emanuel Schnitzer to Langley Research Center associate director, "Space Station Project-results of discussions at NASA Headquarters on May 10 and 11, 1960," May 23, 1960; Emanuel Schnitzer to Langley Research Center associate director, "Transmittal of Updated Article on Erectable Torus Manned Space Laboratory to the American Rocket Society," October 24, 1960, all in A200-4 Files, at Langley Historical Archives, NASA Langley Research Center.

39. The Goodyear Aircraft Company tried to counteract these criticisms in its announcement in the *U.S. Naval Institute Proceedings* that it was prepared to build an inflatable space station (February 1961): 9; and in Darrell C. Romick et. al., "Goodyear Prepares Smaller Space Station," *Aviation Week* 67 (October 14, 1957): 115ff. More recent enthusiasm for an inflatable space station may be found in John C. Mankins, "Technologies for Future Exploration," unpublished presentation to the forty-third American Astronautical Society National Conference, December 10, 1997; Willy Z. Sadeh, Jenine E. Abarbanel, and Marvin E. Criswell, "Inflatable Structures for the

Moon/Mars Surfaces," unpublished presentation to the forty-third American Astronautical Society National Conference, December 10, 1997; Rick W. Sturdevant, "Report on the AAS National Conference, Space Exploration and Development: Beyond the Space Station," *Space Times: Magazine of the American Astronautical Society* 36 (January–February 1997): 11; House Science Committee, Subcommittee on Space and Aeronautics, "NASA Posture Hearing: FY 00 Budget Request," February 24, 1999; House Science Committee, Subcommittee on Space and Aeronautics, "Hearing on FY 2000 Budget Request—Human Space Flight Account," February 25, 1999.

40. Floyd L. Thompson to NASA, Code RSS, "100-foot-diameter automatic erecting modular space station—Request for approval of study contract," June 9, 1961, A200-4, LCF. For design information, see Rene A. Berglund, "Space Station Research Configurations," in NASA Technical Note D-1504, pp. 9–21. The opening paper of Technical Note D-1504, by Paul R. Hill and Emanual Schnitzer, titled "Space Station Objectives and Research Guidelines" (1–9), offers a general statement of early Langley goals and ambitions regarding its space station work as seen in connection to the concepts of both the inflatable torus and rotating hexagon. For North America's presentation of the concept, see North American Aviation, Inc., "Self-Deploying Space Station, Final Report," SID 62-1302, 31 October 1962, copy in the Langley Research Center Technical Library.

41. See James R. Hansen, *Enchanted Rendezvous: John C. Houbolt and the Genesis of the Lunar-Orbit Rendezvous Concept* (Washington, D.C.: Monographs in Aerospace History, no. 4, 1995).

42. This story has been told in John M. Logsdon, "Selecting the Way to the Moon: The Choice of the Lunar Orbital Rendezvous Mode," *Aerospace Historian* 18 (summer 1971): 63–70; Courtney G. Brooks, James M. Grimwood, and Loyd S. Swenson Jr., *Chariots for Apollo: A History of Manned Lunar Spacecraft* (Washington, D.C.: NASA SP-4205, 1979), 61–86; Roger E. Bilstein, *Stages to Saturn: A Technological History of the Apollo/Saturn Launch Vehicle* (Washington, D.C.: NASA SP-4206, 1980), 57–68; Hansen, *Enchanted Rendezvous.*

43. John C. Houbolt, "Lunar Rendezvous," *International Science and Technology* 14 (February 1963): 62–65.

44. George Alexander, "Marshall Intensifies Rendezvous Studies," *Aviation Week* (March 19, 1962): 78–85; *Aviation Week* (January 28, 1963): 27; George von Tiesenhausen, "An Orbital Launch Facility," IAS Paper 63-64, presented at thirty-first annual meeting, January 21–23, 1963; George von Tiesenhausen, "Toward an Orbital Launch Facility," *Astronautics and Aerospace Engineering* (March 1963): 52–55.

45. Sperry Rand Systems Group Staff, "Orbital Launch Operations," vol. 1, prepared for Marshall Space Flight Center, Huntsville, Ala., January 1962, chap. 1, 1; chap. 2, 1–3, A200-4 Files, at Langley Historical Archives, NASA Langley Research Center.

46. "Concluding Remarks by Dr. Wernher von Braun about Mode Selection given to Dr. Joseph F. Shea, Deputy Director (Systems), Office of Manned Space Flight," June 7, 1962, NASA Historical Reference Collection.

47. Arthur C. Clarke, *2001: A Space Odyssey* (New York: New American Library, 1968); Roger D. Launius, "2001: The Odyssey Continues," *Space Times: Magazine of the American Astronautical Society* 40 (January–February 2001): 3, 22.

48. Frederick I. Ordway III, "The Evolution of Space Fiction in Film," in Roger D. Launius, ed., *History of Rocketry and Astronautics: Proceedings of the Fifteenth and Sixteenth History Symposia of the International Academy of Astronautics,* AAS History Series, vol. 11 (San Diego, Calif.: Univelt, 1994), 13–29; Frederick I. Ordway III, "Space Fiction in Film—2001: A Space Odyssey in Retrospect," in Eugene M. Emme, ed., *Science Fiction and Space Futures: Past and Present,* AAS History Series, vol. 5 (San Diego, Calif.: Univelt, 1982); Jerome Agel, ed., *The Making of Kubrick's 2001* (New York: New American Library, 1970).

49. House Science and Technology Committee, Subcommittee on Space Science and Applications, *NASA's Space Station Activities,* 98th Congress, 1st session 1983 (Washington, D.C.: Government Printing Office, 1983), 4.

50. Mitchell Waldrop, "Space City: 2001 It's Not," *Science 83* (October 1983); see also John Noble Wilford, "When Man Has Stations in Space," *New York Times,* October 19, 1969.

51. Sylvia D. Fries, "2001 to 1994: Political Environment and the Design of NASA's Space Station System," in LaFollette and Stine, eds., *Technology and Choice: Readings from Technology and Culture,* 233–58.

3. SKYLAB AND THE SALYUTS: PRELIMINARY SPACE STATIONS

1. The standard work on Skylab is Compton and Benson, *Living and Working in Space.* See also Donald C. Elder, "The Human Touch: The History of the Skylab Program," in Pamela E. Mack, ed., *From Engineering Science to Big Science: The NACA and NASA Collier Trophy Research Project Winners* (Washington, D.C.: NASA SP-4219, 1998), 213–34.

2. On the United States Air Force's Manned Orbiting Laboratory (MOL) concept, see Howard E. McCurdy, *The Space Station Decision: Incremental Politics and Technological Choice* (Baltimore: Johns Hopkins University Press, 1990), 70, 132–33; Walter A. McDougall, *The Heavens and the Earth: A Political History of the Space Age* (New York: Basic Books, 1985), 340–41; Roy F. Houchin III, "Interagency Rivalry: NASA, the Air Force, and MOL," *Quest* 4 (winter 1995): 40–45.

3. One of the best discussions of the Soviet Union's military Salyuts can be found in Phillip Clark, *The Soviet Manned Space Programme: An Illustrated History of the Men, the Missions and the Spacecraft* (London: Salamander Books, 1988), 66–75. See also Dennis Newkirk, *Almanac of Soviet Manned Space Flight* (Hous-

ton, Tex.: Gulf Publishing Company, 1990), 112–53; Wayne R. Matson, *Cosmonautics: A Colorful History* (Washington, D.C.: Cosmos Books, 1994), 52–75.

4. The classic study of this subject is Leon Festinger, *When Prophesy Fails* (Minneapolis: University of Minneapolis Press, 1956). I have also found this in my studies of Mormon history. See my "Quest for Zion: Joseph Smith III and Community-Building in the Reorganization, 1860–1900," in Maurice L. Draper and Debra Combs. eds., *Restoration Studies III* (Independence, Mo.: Herald Publishing House, 1986), 314–32.

5. Floyd L. Thompson to distribution, with replies, July 13, 1962; George W. Brooks to Thompson, "Faget's Remarks during Symposium of 7/31/62-8/1/62," August 7, 1962, both in A200-4 Files, at Langley Historical Archives, NASA Langley Research Center.

6. For a more detailed examination of some of the difficulties NASA managers faced in attempting to agree on a set of proposed missions for a space station program, see Gruen, "The Port Unknown," chap. 1.

7. Hook, "Space Stations," 4.

8. The standard work on the Gemini program is Barton C. Hacker and James M. Grimwood, *On the Shoulders of Titans: A History of Project Gemini* (Washington, D.C.: NASA SP-4203, 1977).

9. Hook, "Space Stations," 4; see also the description of the MORL concept in the NASA Langley press release, "Industry Asked to Propose Manned Orbiting Research Laboratory Plans," April 24, 1963, NASA Historical Reference Collection.

10. Floyd L. Thompson, Langley Research Center Announcement No. 32-63, "Reorganization of Langley Research Center and reassignment of personnel effective June 10, 1963," June 6, 1963; Floyd L. Thompson, Langley Research Center Announcement No. 33-63, "Changes in Organization of the Applied Materials and Physics Division," June 6, 1963. Both in Floyd L. Thompson collection, Langley Historical Archives, NASA Langley Research Center.

11. Phase I in aerospace engineering is the preliminary analysis and definition effort for any project. For MORL this had been completed by Langley in the summer of 1963. Phase II is the preliminary design and development work and usually is awarded by NASA to an industry partner. Phases III and IV include the detailed design and development work leading to the actual building of hardware.

12. Hook, "Space Stations," 4.

13. William N. Gardner, head, MORL Studies Office, to Charles Donlan, "MORL Studies Office Weekly Report—Week ending June 22, 1963," June 24, 1963, A200-4 Files, at Langley Historical Archives, NASA Langley Research Center.

14. William Nissim, Douglas Aircraft Company, "London Daily Mail Astronomical Space Observatory," Douglas Report No. SM-36173, November 1959, copy in Langley Research Center Technical Library; C.C. Walkey, Contracts Division, Douglas Aircraft Company, to Laurence K. Loftin Jr., "Unsolicited Proposal for Advanced Technology Studies Related to Orbiting Laboratories," July 13, 1964, A200-4 Files, at Langley Historical Archives, NASA Langley Research Center.

15. John W. Massey, "Historical Resume of Manned Space Stations," Army Ballistic Missile Agency, Redstone Arsenal, Ala., Report No. DSP-TM-9-60, June 15, 1960, NASA Historical Reference Collection; McCurdy, *Space Station Decision,* 101–5, 131–34, 149–50, 166–68, 198–99.

16. Douglas Aircraft Company, Missile and Space Systems Division, "MORL: Manned Orbital Research Laboratory," Douglas Report SM-47966, August 1964; "Report on the Optimization of the Manned Orbital Research Laboratory (MORL) System Concept," Douglas Report SM-46071, September 1964; "Report on the Development of the Manned Orbital Research Laboratory (MORL) System Utilization Potential," Douglas Report SM-48822, January 1966. All available in Langley Research Center Technical Library.

17. Robert S. Osborne, Space Station Research Group, to Charles Donlan, associate director, Langley Research Center, "OART Meeting on MOL Uses," November 27, 1963; Maurice J. Raffensperger, director, Manned Earth Orbital Missions Studies, to distribution, "Minutes of the Orbital Research Laboratory Study Reporting Meeting No. 12," June 1964, 7, 9. Both in Container 047-2, Johnson Space Center History Office Archives, Houston, Tex. See also Robert N. Parker, Manned Orbital Research Laboratory Studies Office, to Charles Donlan, associate director, Langley Research Center, March 10, 1964, A200-4 Files, at Langley Historical Archives, NASA Langley Research Center; Michael I. Yarymovych, acting director, Manned Earth Orbital Mission Studies, to W. N. Gardner, "Suspension of the Study of Military Uses of the MORL," March 10, 1964, NASA Historical Reference Collection.

18. Edward Olling, "Project Olympus, S-140 NASA-MSC," July 16, 1962, Johnson Space Center History Office Archives, Houston, Tex.

19. Douglas R. Lord and Emanuel Schnizter to Joseph F. Shea, "Comments and Recommendations on a Proposed Summary Project Development Plan (PDP) for a Space Station Program," October 15, 1962; Douglas R. Lord to Shea and William A. Lee, "Summary of Minutes of Manned Space Station Meeting, 9/28/62," October 2, 1962, 2. Both in NASA Historical Reference Collection. Douglas R. Lord to Max Faget, "Comments on Contractual Study Proposed by MSC Entitled 'Operations and Logistics Study of Manned Orbiting Space Station,'" November 15, 1962, Folder NAS9-1422, Box 8, Federal Retired Records Container 70A5404 (Project Files Code MF), Suitland, Md.

20. The von Braun quote is from S. Fred Singer, ed., *Manned Laboratories in Space* (New York: Springer-Verlag, 1969), 1.

21. Douglas Aircraft Company and International Business Machines Corporation, Douglas Missile and Space Systems Division, "Final Report on MOL (Zero G) Saturn V Class," vol. 1 and Condensed Summary (SM-45560), March 1964, 1–8, 14, 19, 22–25, Boxes 3 and 8, Federal Retired Records Container 70A5404, "OMSF Project Files," Suitland, Md.; Lockheed-California Company, Spacecraft Organization, "Study of a Rotating Manned Orbital Space Station," Final Report, vol. 11, Summary, LR 17502, and vol. 1, Summary and Conclusions, Sections 1 through 6, LR

17366, November 30, 1963, both in Box 7, Federal Retired Records Container 70A5404, "OMSF Project Files," Suitland, Md.; North American Aviation, Space and Information Systems Division, "Extended-Mission Apollo Study, Final Report, Condensed Summary and Addendum," vols. 2 through 11 and Final Technical Review, NAA pub. no. SID 63-1370-12, pp. 2, 4, 13, 16, Box 6, Federal Retired Records Access Number 70A5404, "OMSF Project Files," Suitland, Md.; McDonnell Aircraft Corporation, "Modular Space Station Evolving from Gemini, Report No. 9272," vols. 1 and 2, December 15, 1962, Johnson Space Center History Office Archives, Houston, Tex.

22. Douglas, "MORL," August 1964, 3, 22; North American Aviation, Space and Information Systems Division, "Extended-Mission Apollo Study, Final Report, Condensed Summary and Addendum," vols. 2 through 11 and Final Technical Review, NAA pub. no. SID 63-1370-12; North American Aviation, Space and Information Systems Division, "Modular Concepts Investigation Interim Report," pub. no. SID 63-1516, December 16, 1963. Both in Box 6, Federal Retired Records Access Number 70A5404, "OMSF Project Files," Suitland, Md.

23. A piloted Gemini spacecraft would not fly until March 23, 1965; the first piloted flight involving rendezvous did not take place until June 3, 1965; and the first piloted flight involving docking (the joining of two spacecraft to become one) took place only on March 16, 1966. See Hacker and Grimwood, *On the Shoulders of Titans,* Appendix B, "Flight Data on Gemini III and IV." The "Tinker Toy" label is attributed to Edward Olling of the Manned Spacecraft Center. See "NASA Favors Deployable Stations for First-Generation Space Study," *Aviation Week and Space Technology* (April 15, 1963): 61.

24. "NASA Building Space Station Technology," *Aviation Week and Space Technology* (July 22, 1963): 80.

25. This has been discussed in detail in Mark A. Erikson, "The Evolution of the NASA-DoD Relationship from Sputnik to the Lunar Landing." Ph.D. diss., George Washington University, 1997.

26. "Air Force Given Space Laboratory Mission," *Aviation Week and Space Technology* (December 16, 1963): 29–30; "Agreement Reached on MOL Status, NASA to Continue Station Studies," *Aviation Week and Space Technology* (January 6, 1964): 28; William Leavitt, "MOL: Evolution of a Decision," *Air Force Magazine* 48 (October 1965): 37.

27. William Leavitt, "Memo on MOL," *Air Force Magazine* 47 (February 1964): 74–76; Robert C. Seamans Jr. and Harold Brown, memorandum for the Secretary of Defense and the NASA administrator, "Joint NASA-DOD Review of National Space Program," January 31, 1964, NASA-DOD Coordination File, NASA Historical Reference Collection. A more detailed overview of the agreement was presented in Harold Brown, director of Defense Research and Engineering, to George P. Miller, chairman, House Committee on Science and Astronautics, "NASA-DOD Relationships," March 6, 1964, in U.S. House of Representatives, *Report of the Subcommittee on NASA Oversight of the Committee on Science and Astronautics, The NASA-DOD Relationship,* 88th Congress, 2d session (Washington, D.C.: Government Printing Office, 1964), 8.

28. This paragraph and the following ones on MOL are based on Donald Pealer, "Manned Orbiting Laboratory (MOL), Part 1," *Quest* 4 (fall 1995): 4–17; Donald Pealer, "Manned Orbiting Laboratory (MOL), Part 2," *Quest* 4 (winter 1995): 28–37; Donald Pealer, "Manned Orbiting Laboratory (MOL), Part 3," *Quest* 5, No. 2 (1996): 16–23.

29. On Truly, see John M. Logsdon, "Return to Flight: Richard H. Truly and the Recovery from the *Challenger* Accident," in Pamela E. Mack, ed., *From Engineering Science to Big Science: The NACA and NASA Collier Trophy Research Project Winners* (Washington, D.C.: NASA SP-4219), 1998), 345–64.

30. Compton and Benson, *Living and Working in Space,* 18, 40–56; W. David Compton, "The Rocket as Spacecraft: Spent Stages in Manned Space Flight," *Journal of the British Interplanetary Society,* 38 (1985): 147–54.

31. Leland F. Belew with Ernst Stuhlinger, *Skylab: A Guidebook* (Washington, D.C.: NASA EP-107, 1973), 6–7.

32. See John A. Eddy, *A New Sun: The Solar Results from Skylab* (Washington, D.C.: NASA SP-402, 1979). The book describes the Apollo Telescope Mount, the role of the scientist-astronauts, and the results.

33. Linda Neumann Ezell, compiler, *NASA Historical Data Book,* vol. 3, *Programs and Projects, 1969–1978* (Washington, D.C.: NASA SP-4012, 1988), 94–98.

34. *Skylab Experiments,* vol. 5, *Astronomy and Space Physics* (Washington, D.C.: NASA EP-114, 1973).

35. Richard S. Johnston and Lawrence F. Dietlein, eds., *Biomedical Results from Skylab* (Washington, D.C.: NASA SP-377, 1977), 3–16; Leland F. Belew, *Skylab: Our First Space Station* (Washington, D.C.: NASA SP-400, 1977), chap. 1; Arnauld E. Nicogossian, Carolyn Leach Huntoon, and Sam L. Pool, *Space Physiology and Medicine,* 3d ed. (Philadelphia, Pa.: Lea and Wiger, 1994), 10–12.

36. Bruce T. Lundin et al., "NASA Investigation Board Report on the Initial Flight Anomalies of Skylab 1," July 13, 1973, NASA Historical Reference Collection, and available on-line at http://history.nasa.gov/skylabrep/SRcover.htm, accessed January 19, 2002.

37. Compton and Benson, *Living and Working in Space,* 104–11; George E. Mueller to administrator, "Saturn V Launched Workshop for AAP," July 10, 1969; William C. Schneider to Robert F. Thompson, "AAP PAD Change Request," July 29, 1969, with attachments, both in NASA Historical Reference Collection.

38. Belew, *Skylab,* 14.

39. Ibid., 62–68.

40. Helen T. Wells, Susan H. Whiteley, and Carrie Karegeannes, *Origins of NASA Names* (Washington, D.C.: NASA SP-4402, 1976), chap. 3.

41. Richard Nixon, memorandum for the vice president, Secretary of Defense, acting administrator NASA, and Science Adviser, February 13, 1969, NASA Historical Reference Collection.

42. Thomas O. Paine, acting administrator, NASA, memorandum for the president, "Problems and Opportunities in Manned Space Flight," February 26, 1969, NASA Historical Reference Collection.

43. Robert C. Seamans Jr., Secretary of the Air Force, to Honorable Spiro T. Agnew, vice president, August 4, 1969, NASA Historical Reference Collection.

44. President's Space Task Group, *The Post-Apollo Space Program: Directions for the Future* (Washington, D.C.: Executive Office of the President, September 1969); *New York Times,* September 16, 1969, 1; D.E. Crabill, Bureau of the Budget, to director, Bureau of the Budget, "President's Task Group on Space—Meeting No. 2," March 14, 1969, Record Group 51, Series 69.1, Box 51-78-31, National Archives and Records Administration, Washington, D.C.; Clay T. Whitehead, staff assistant White House, to Peter M. Flanigan, White House, June 25, 1969, Record Group 51, Series 69.1, Box 51-78-31, National Archives; Robert P. Mayo, director, Bureau of the Budget, to President Richard M. Nixon, "Space Task Group Report," September 25, 1969, Record Group 51, Series 69.1, Box 51-78-31, National Archives; John Erlichman, *Witness to Power: The Nixon Years* (New York: Pocket Books, 1982), 123–24.

45. White House press secretary, "The White House, Statement by the President," March 7, 1970, Presidential Files, NASA Historical Reference Collection.

46. Lundin et al., "NASA Investigation Board Report on the Initial Flight Anomalies of Skylab 1," Summary, July 13, 1973.

47. Dwayne A. Day, "The Air Force in Space: Past, Present and Future," *Space Times: Magazine of the American Astronautical Society* 35 (March–April 1996): 17.

48. Compton and Benson, *Living and Working in Space,* 269–71; Robert Kain, Crew Training and Procedures Division, NASA Johnson Space Center, "Skylab EVA," n.d.; NASA Johnson Space Center, *Skylab Experience Bulletin No. 27: Personnel and Equipment Restraint and Mobility Aids: EVA,* Johnson Space Center 09561, May 1975, NASA Johnson Space Center Library, Houston, Tex.; David S. F. Portree and Robert C. Treviño, compilers, *Walking to Olympus: A Chronology of Extravehicular Activity (EVA)* (Washington, D.C.: Monographs in Aerospace History, no. 7, 1997), 30–31.

49. Portree and Treviño, *Walking to Olympus,* 30–31; Compton and Benson, *Living and Working in Space,* 269–71; "The Skylab Missions," *Marshall Star,* May 11, 1988; "Record Payload for Next Skylab," *Aviation Week and Space Technology* (July 2, 1973): 16.

50. NASA Life Sciences Data Archive, "Skylab 2," available on-line at http://lsda.jsc.nasa.gov/skylab/skylab2.stm, accessed January 20, 2002.

51. Compton and Benson, *Living and Working in Space,* 339–54.

52. Interview with Senator Jake Garn, by Jeffrey Bingham, May 18, 2000, NASA Historical Reference Collection.

53. Nicogossian, Huntoon, and Pool, ed., *Space Physiology and Medicine,* 12.

54. Ibid.

55. James V. Zimmerman for Arnold W. Frutkin, assistant administrator for International Affairs, to Dr. John V. N. Granger, acting director, Bureau of International Scientific and Technological Affairs, Department of State, September 12, 1974, with attached: "Foreign Policy Issues Regarding Earth Resource Surveying by Satellite: A Report of the Secretary's Advisory Committee on Science and Foreign Affairs," July 24, 1974, NASA Historical Reference Collection.

56. "Feasibility Study of Commercial Space Manufacturing, Phase II Final Report," vol. 1: Executive Summary, McDonnell Douglas Astronautics Company, East, St. Louis, Mo., January 15, 1977, MDC E1625, NASA Historical Reference Collection.

57. Paul D. Spudis, "Robots vs. Humans: Who Should Explore Space," *Scientific American* 10 (spring 1999): 25, 30–31 (quote from 31).

58. This section on the Soviet space stations is based on the Salyut files in the NASA Historical Reference Collection, as well as on Matson, *Cosmonautics,* 50–55, 60–75; Caprara, *Living in Space,* 62–107; Saunders B. Kramer, "The Rescue of Salyut 7," *Air and Space* 4 (November/December 1990): 54–59; Daniel James Gauthier, "Salute to Salyut: A History of Soviet Station Programs." *Quest* 1, no. 1 (spring 1992): 16–23; Asif A. Siddiqi, "The Triumph and Tragedy of Salyut 1," *Quest* 5, no. 3 (1996): 26–3.

59. I. B. Afanasiyev, *Novoye v Zhizni, Nauke, Tekhnike: Seriya Kosmonavtika, Astronomiya* (Unknown spacecraft), vol. 12 (Moscow: Nauka, 1991).

60. Robert Zimmerman, *The Chronological Encyclopedia of Discoveries in Space* (Phoenix, Ariz.: Ornx Press, 2000), 95.

61. Nicogossian, Huntoon, and Pool, eds., *Space Physiology and Medicine,* 21–26.

62. Day, "The von Braun Paradigm," 12–15 (quote from 15).

4. The Strange Career of Space Station Freedom

1. Hook, "Space Stations"; Theodore R. Simpson, ed., *The Space Station: An Idea Whose Time Has Come* (Washington, D.C.: IEEE Press, 1985), 119–36.

2. *Outlook for Space* (Washington, D.C.: Government Printing Office, 1976), 200.

3. James M. Beggs, "Why the United States Needs a Space Station," remarks prepared for delivery at the Detroit Economic Club and Detroit Engineering Society, June 23, 1982, NASA Historical Reference Collection, reprinted under the same title in *Vital Speeches* 48 (August 1, 1982): 615–17.

4. McCurdy, *Space Station Decision,* 40.

5. George M. Low, team leader, NASA Transition Team, to Richard Fairbanks, director, Transition Resources and Develop-

ment Group, December 19, 1980, with attached: "Report of the Transition Team, National Aeronautics and Space Administration," George M. Low Papers, Institute Archives and Special Collections, Rensselaer Polytechnic Institute, Troy, N.Y.

6. Hans Mark and Milton Silveira, "Notes on Long Range Planning," August 1981, NASA Historical Reference Collection.

7. For a discussion of this strategy see Hans Mark, *The Space Station: A Personal Journey* (Durham, N.C.: Duke University Press, 1987), 162–87.

8. National Security Decision Directive 5-83, "Space Station," April 11, 1983, National Security Archive, Washington, D.C.

9. Interagency Group (Space), "Manned Space Station Study," enclosure in Gen. Richard G. Stilwell, Deputy Under Secretary of Defense, to Robert C. McFarlane, June 20, 1983, Department of Defense folder, Space Station Task Force files, 1982–1984, Record Group 255, Federal Records Center, National Archives and Records Administration, Suitland, Md.

10. McCurdy, *Space Station Decision,* 177–85.

11. "Revised Talking Points for the Space Station Presentation to the President and the Cabinet Council," November 30, 1983, with attached: "Presentation on Space Station," December 1, 1983, NASA Historical Reference Collection.

12. Donald R. Baucom, *The Origins of SDI, 1944–1983* (Lawrence: University Press of Kansas, 1992), 171–200; Ronald Reagan, *Ronald Reagan: An American Life* (New York: Simon and Schuster, 1990), 547–48, 601–9, 650–54, 676–79; Roger Handberg, *Seeking New World Vistas: The Militarization of Space* (Westport, Conn.: Praeger, 2000), 65–85.

13. "State of the Union Message, January 25, 1984," *Public Papers of the Presidents of the United States: Ronald Reagan, 1984* (Washington, D.C.: Government Printing Office, 1986), 87–95.

14. Caspar Weinberger, Secretary of Defense, to James M. Beggs, NASA administrator, January 16, 1984, NASA Historical Reference Collection.

15. McCurdy, *Space Station Decision,* 80–98.

16. NASA Technical Memorandum TM—86652, "Space Station Program Description Document," books 1 through 7, final edition, March 1984, NASA Historical Reference Collection.

17. *Defense Daily,* March 3, March 14, July 19, and August 3, 1983; *Aerospace Daily,* March 1, June 3, and July 19, 1983.

18. Quoted in McCurdy, *Space Station Decision,* 171.

19. "Pie in the Sky: Big Science Is Ready for Blastoff," *Congressional Quarterly Weekly Report,* April 28, 1990, 1254–59; "Overhaul Ordered for Space Station Design," *Congressional Quarterly Almanac 1990* (Washington, D.C.: Congressional Quarterly, 1991), 436.

20. Wernher von Braun, "Our Future in Space," NASA Marshall Space Flight Center Public Information Office, Huntsville, Ala., February 1, 1962, 7.

21. U.S. Congress, Office of Technology Assessment, *Proceedings of the Workshop on Automation and Space Station, 1984,* published in U.S. Senate, Subcommittee of the Senate Commit-

tee on Appropriations, *Hearings on Housing and Urban Development, and Certain Independent Agencies Appropriations for Fiscal Year 1985,* 98th Congress, 2d session, March 29, 1984 (Washington, D.C.: Government Printing Office, 1984), 1266, 1289–90.

22. Sylvia D. Fries, "2001 to 1994: Political Environment and the Design of NASA's Space Station System," in LaFollette and Stine, eds., *Technology and Choice: Readings from Technology and Culture,* 253–57.

23. A NASA Concept Development Group met throughout 1983 to develop a set of possible design concepts.

24. Bass Redd, "Space Station Reference Configuration for RFP Level B Presentation to Level A," May 17, 1984, 21, included in NASA Space Station Program Director's Meeting Presentation Materials, May 17, 1984, NASA Historical Reference Collection.

25. Neil B. Hutchinson, "Space Station Definition Phase Industry Prepoposal Conference," presentation materials, subsection "Requirements Methodology," September 25, 1984, NASA Historical Reference Collection.

26. Thomas G. Mancuso to distribution, "Charts for Space Station Industry Briefing Conducted 7-11-84," July 23, 1984; NASA, "Space Station Definition and Preliminary Design Request For Proposal," September 15, 1984; Neil Hutchinson, "Space Station Definition Phase Industry Prepoposal Conference," September 25, 1984; NASA, "Space Station Definition and Preliminary Design Request For Proposal," September 15, 1984; NASA, "Space Station Definition and Preliminary Design Request For Proposal," 15 September 1984; Neil Hutchinson to Philip Culbertson, May 6, 1985; Carolyn Townsend to distribution, June 14, 1985; James M. Beggs to Philip Culbertson, "Selection of Contractors for Space Station Definition and Preliminary Design," March 15, 1985. All in NASA Historical Reference Collection.

27. NASA Office of Space Station, "Space Station Management Plan and Procurement Strategy," submitted to the Science and Technology Committee of the U.S. House of Representatives, December 14, 1984; NASA Office of Space Station, "Monthly Briefing," presentation charts, June 1985. Both in NASA Historical Reference Collection.

28. Gruen, "The Port Unknown," chap. 5, p. 17.

29. Charles N. Crews to distribution, "Minutes for the Level B SSCB Meeting," July 1, 1985; Crews to distribution, "Minutes and Directives of February 20–21, 1986, Level B SSCB Meeting," May 2, 1986; Crews to distribution, "Minutes for Level B SSCB," August 5, 1985; Crews to distribution, "Minutes for Level B SSCB," August 29, 1985; Crews to distribution, "Minutes of September 12, Level B Space Station Control Board Meeting," October 2, 1985; Crews to distribution, "Minutes of October 17 [1985] Level B Space Station Control Board Meeting," November 5, 1985. All in NASA Historical Reference Collection.

30. Crews to distribution, "Minutes for Level B SSCB," August 5, 1985; Crews to distribution, "Minutes for Level B Space Station Control Board (SSCB) Meeting," September 24,

1985. Both in NASA Historical Reference Collection.

31. Crews to distribution, "Minutes of October 3 Level B Space Station Control Board Meeting," October 13, 1985; "Minutes and Directives of October 31, 1985, Level B SSCB Meeting," November 20, 1985. Both in NASA Historical Reference Collection.

32. Richard W. Hautamaki, executive secretary, Level B Space Station Systems Integration Board (SIB) to distribution, "Minutes of October 9–10, 1985, Level B Space Station SIB Meeting," October 22, 1985; Crews to distribution, "Minutes of October 17 Level B Space Station Control Board Meeting," November 5, 1985. Both in NASA Historical Reference Collection.

33. McCurdy, *Space Station Decision*, 223–35.

34. Walter J. Oleszek, *Congressional Procedures and the Policy Process*, 4th ed. (Washington, D.C.: Congressional Quarterly, 1996), 21–22.

35. "*Freedom* Fighters Win Again: Senate Keeps Space Station," *Congressional Quarterly Weekly Report*, September 12, 1992, 2722.

36. W. Henry Lambright, *Powering Apollo: James E. Webb of NASA* (Baltimore: Johns Hopkins University Press, 1995), 132–41; Robert Dallek, "Johnson, Project Apollo, and the Politics of Space Program Planning," in Roger D. Launius and Howard E. McCurdy, eds., *Spaceflight and the Myth of Presidential Leadership* (Urbana: University of Illinois Press, 1997), 79–88.

37. Lyn Ragsdale, "Politics Not Science: The U.S. Space Program in the Reagan and Bush Years," in Launius and McCurdy, eds., *Spaceflight and the Myth of Presidential Leadership*, 133–71; Jeffrey M. Bingham, "A New Task for Space Policy History: Understanding Congress and the International Space Station," unpublished paper in possession of author, July 22, 1999.

38. "Pie in the Sky: Big Science Is Ready for Blastoff," 1254–59; "Overhaul Ordered for Space Station Design," *Congressional Quarterly Almanac 1990* (Washington, D.C.: Congressional Quarterly, 1991), 436; Marcia S. Smith, "NASA's Space Station Program: Evolution and Current Status," House Science Committee testimony, April 4, 2001, NASA Historical Reference Collection.

39. Andrew Stofan and Thomas Moser, "Space Station Program Cost Assessment to NASA Administrator," January 22, 1987; Andrew Stofan, Thomas Moser, and John Aaron, "Space Station Program Cost Assessment to NASA Administrator," January 29, 1987; James C. Fletcher to the president, January 30, 1987. All in NASA Historical Reference Collection.

40. Marcia S. Smith, "NASA's Space Station Program: Evolution and Current Status," House Science Committee testimony, April 4, 2001, NASA Historical Reference Collection.

41. John M. Logsdon, "International Cooperation in the Space Station Programme: Assessing the Experience to Date," *Space Policy* 7 (February 1991): 35–36; John M. Logsdon, *Together in Orbit: The Origins of International Participation in Space Station Freedom* (Washington, D.C.: Monographs in Aerospace History, no. 11, 1998).

42. McCurdy, *Space Station Decision*, 101–2; U.S. Congress, Office of Technology Assessment, *International Partnerships in Large Science Projects,* OTA-BP-ETI-150 (Washington, D.C.: Government Printing Office, 1995).

43. Logsdon, "International Cooperation in the Space Station Programme," 35–45; Ernst W. Messerschmid, "A European Perspective: International Cooperation for Future Space Science Missions," *Space Times: Magazine of the American Astronautical Society* 34 (May–June 1995): 2–6.

44. Quoted in McCurdy, *Space Station Decision,* 194.

45. Adam L. Gruen, "Deep Space Nein? The Troubled History of Space Station *Freedom,*" *Ad Astra* (May–June 1993): 18–23. Copy of space station cartoon undated, in possession of author.

46. Smith, "NASA's Space Station Program."

47. "Mikulski Trims Carefully to Pay Cost of Freedom," *Congressional Quarterly Weekly Report,* July 13, 1991, 1890.

48. Private comment of senior Space Station Freedom manager to author, January 14, 1992.

49. John Law, "Technology and Heterogeneous Engineering: The Case of Portuguese Expansion," 111–34; Donald MacKenzie, "Missile Accuracy: A Case Study in the Social Processes of Technological Change," 195–222. Both in Wiebe E. Bijker, Thomas P. Hughes, and Trevor J. Pinch, eds., *The Social Construction of Technological Systems: New Directions in the Sociology and History of Technology* (Cambridge: MIT Press, 1987).

5. A MIR INTERLUDE

1. Zimmerman, *The Chronological Encyclopedia of Discoveries in Space,* 219–20; Bart Hendrickx, "The Origins and Evolution of Mir and Its Modules," *Journal of the British Interplanetary Society* 51 (June 1998): 203–22.

2. *Aeronautics and Space Report of the President, 1999 Activities* (Washington, D.C.: NASA Annual Report, October 2001), Appendix C.

3. Nicogossian, Huntoon, and Pool, eds., *Space Physiology and Medicine,* 26; David J. Shayler, "Outside Mir: Ten Years of EVA Operations," *Journal of the British Interplanetary Society* 51 (January 1998): 29–38.

4. Clay Morgan, *Shuttle-Mir: The U.S. and Russia Share History's Highest Stage* (Washington, D.C.: NASA SP-2001-4225, 2001), 164.

5. On the general history of Mir, see David M. Harland, *The Mir Space Station: A Precursor to Space Colonization* (New York: John Wiley and Sons, 1997); Lee Robert Caldwell, "Twelve Years of Mir Space Station Operations," *Journal of the British Interplanetary Society* 50 (August 1997): 317–20; Andy L. Salmon, "Science Onboard the Mir Space Station, 1984–1994," *Journal of the British Interplanetary Society* 50 (August 1997): 283–95; Neal Bernards, *Mir Space Station* (New York: Smart Apple Media, 1999).

6. E-mail from Trish Graboske to author, "Nicklas's Law," February 21, 2002, copy in possession of author.

7. This characterization provided the title for Bryan Burrough, *Dragonfly: NASA and the Crisis aboard the Mir* (New York: Ballinger, 1998).

8. Jerry M. Linenger, *Off the Planet: Surviving Five Perilous Months aboard the Space Station Mir* (New York: McGraw-Hill, 1999), 95.

9. Michael Foale oral history, June 16, 1998, quoted in Morgan, *Shuttle-Mir,* 163.

10. U.S. Congress, Office of Technology Assessment, *U.S.-Soviet Cooperation in Space,* OTA-TM-STI-27 (Washington, D.C.: Government Printing Office, July 1985), 5.

11. Clay Morgan, "Competition, Cooperation, and Compromise: An Annotated Chronology of Shuttle-Mir and Its Background," 2001, 11, unpublished paper in possession of author.

12. Office of the Press Secretary, the White House, "Joint Statement on Cooperation in Space," June 17, 1992, NASA Historical Reference Collection.

13. "Implementing Agreement between the National Aeronautics and Space Administration of the United States of America and the Russian Space Agency of the Russian Federation on Human Spaceflight Cooperation," October 5, 1992; Office of the Vice President, the White House, "United States–Russian Joint Commission on Energy and Space—Joint Statement on Cooperation in Space," September 2, 1993, both in NASA Historical Reference Collection.

14. Shuttle flight STS 60 in February 1994, involving the participation of cosmonaut Sergei Krikalev in a Space Shuttle mission, was formally also considered part of the Phase One program. A Shuttle manifest dated November 30, 1994, showed a total of forty-four flights in the four-year construction period. Twenty-seven were to be Space Shuttle flights. Those totals did not include flights to rotate crews at the station or to resupply fuel and other consumables.

15. Quoted from "Protocol to the Implementing Agreement between the National Aeronautics and Space Administration of the United States of America and the Russian Space Agency of the Russian Federation on Human Spaceflight Cooperation," December 16, 1993, NASA Historical Reference Collection.

16. U.S. Congress, Office of Technology Assessment, *U.S.-Russian Cooperation in Space,* OTA-ISS-618 (Washington, D.C.: Government Printing Office, April 1995), 11.

17. Arnold W. Frutkin, *International Cooperation in Space* (Englewood Cliffs, N.J.: Prentice-Hall, 1965), 73, 78; McCurdy, *Space Station Decision,* 101.

18. U.S. Congress, *U.S.-Russian Cooperation in Space,* OTA-ISS-618, 8.

19. Morgan, *Shuttle-Mir,* 8–31.

20. Norman E. Thagard oral history, September 16, 1998, NASA Historical Reference Collection.

21. Ibid. See also Glen E. Swanson, "An Interview with Astronaut Norman Thagard," *Space Times: Magazine of the American Astronautical Society* 34 (September–October 1995): 9–12.

22. This and the following paragraphs about the Atlantis-Mir docking are taken from Roger D. Launius, "Making History in Space, Pointing Directions for the Future: A Review of the Recent Atlantis/Mir Docking Mission," *Space Times: Magazine of the American Astronautical Society* 34 (September/October 1995): 4–8.

23. Some of these same thoughts are expressed in Kenneth T. Jackson's introduction, "The Shape of Things to Come: Urban Growth in the South and West," in Robert B. Fairbanks and Kathleen Underwood, eds., *Essays on Sunbelt Cities and Recent Urban America* (College Station: Texas A&M University Press, 1990), 7.

24. Shannon Lucid oral history, June 17, 1998, NASA Historical Reference Collection.

25. Burrough, *Dragonfly,* 133–34, 145–47, 153–54, 161–66, 180–81.

26. For an autobiographical account of the episode see Linenger, *Off the Planet.*

27. NASA Johnson Space Center press release, "Small Fire Extinguished on Mir," February 24, 1997, NASA Historical Reference Collection.

28. Quoted in Morgan, *Shuttle-Mir,* 92.

29. Ibid.

30. "Small Fire Extinguished on Mir."

31. Frank Culbertson, NASA Shuttle-Mir program director, "Weekly Update," February 28, 1997, NASA Historical Reference Collection.

32. Personal conversation with senior space station official, April 14, 2000.

33. NASA Johnson Space Center press release, "NASA-4/Mir-23 Mission Status Reports, Week of April 25, 1997," NASA Historical Reference Collection.

34. See Linenger, *Off the Planet.*

35. The story of Apollo 13 is engagingly told by James A. Lovell and Jeffrey Kluger in *Lost Moon: The Perilous Voyage of Apollo 13* (New York: Houghton Mifflin, 1994).

36. Quoted in Morgan, *Shuttle-Mir,* 109.

37. Ibid.

38. NASA Johnson Space Center press release, "NASA-4/Mir-23 Mission Status Reports, Week of June 27, 1997," NASA Historical Reference Collection.

39. NASA Johnson Space Center press releasees, "NASA-4/Mir-23 Mission Status Reports," weeks of July 4, 11, 18, 25, and August 1, 8, 15, 22, 29, 1997, NASA Historical Reference Collection.

40. Morgan, *Shuttle-Mir,* 109–14.

41. Statement of Chairman F. James Sensenbrenner Jr., Science Committee, U.S. House of Representatives, June 25, 1997, NASA Historical Reference Collection.

42. Both quoted in Morgan, *Shuttle-Mir,* 159.

43. Ibid.

44. Ibid.

45. Ibid., 158.

46. "A *Mir* Interlude," *Space Times: Magazine of the American Astronautical Society* 40 (May–June 2001): 18.

47. NASA and NCS, "Questions and Answers" on Mir deorbit, February 27, 2001, unpublished paper in possession of author.

48. "After 86,331 Orbits, Mir Space Station's 15 Years in Space Ends: February 19, 1986–2001," March 28, 2001, available on-line at http://www.satobs.org/mir.html, accessed March 10, 2002; State Department press release, "Mir Space Station Deorbit," March 2, 2001, copy in possession of author, and available on-line at http://www.state.gov/r/pa/prs/ps/2001/1035.htm, accessed March 10, 2002.

49. Vladimir Semyachkin oral history, July 17, 1998, quoted in Morgan, *Shuttle-Mir,* 165.

6. BUILDING THE INTERNATIONAL SPACE STATION

1. See Launius and McCurdy, *Imagining Space,* 94–98; W. D. Kay, "Democracy and Super Technologies: The Politics of the Space Shuttle and Space Station Freedom," *Science, Technology, and Human Values* 19 (spring 1994): 131–51.

2. See McCurdy, *Space Station Decision;* Thomas J. Lewin and V. K. Narayanan, *Keeping the Dream Alive: Managing the Space Station Program, 1982–1986* (Washington, D.C.: Contractor Report 4272, 1990); Gruen, "The Port Unknown."

3. On the early international participation in the space station program, see Logsdon, *Together in Orbit.*

4. James Asker, "Space Station Key to NASA's Future," *Aviation Week and Space Technology* (March 15, 1993): 83; NASA press release 93-64, "Gibbons Outlines Space Station Redesign Guidance," April 6, 1993, NASA Historical Reference Collection.

5. NASA press release 93-104, "Station Redesign Team to Submit Final Report," June 4, 1993, NASA Historical Reference Collection.

6. Ibid.

7. Statement by the president, June 17, 1993, NASA Historical Reference Collection; Brian D. O'Connor et al., *Space Station Redesign Team Final Report to the Advisory Committee on the Redesign of the Space Station* (Washington, D.C.: NASA, June 1993).

8. NASA Advisory Council, "Report of the Cost Assessment and Validation Task Force on the International Space Station," April 21, 1998, 4, NASA Historical Reference Collection.

9. Morgan, *Shuttle-Mir,* 71, 134, 159, 161; Maxim V. Tarasenko, "Transformation of the Soviet Space Program after the Cold War," *Science and Global Security* 4 (1994): 339–61.

10. Advisory Committee on the Future of the U.S. Space Program, *Report of the Advisory Committee on the Future of the U.S. Space Program* (Washington, D.C.: Government Printing Office, December 1990), 8; Kenneth S. Pederson, "Thoughts on International Space Cooperation and Interests in the Post–Cold War World," *Space Policy* 8 (August 1992): 216–18.

11. Although the Canadian Mobile Servicing System has been on the station's critical path from the beginning, the agreement provides for all Canadian hardware, plans, and materials to be turned over to NASA in the event Canada should withdraw from the program. As in the agreement for the Shuttle's Canadarm, this gives the agency ultimate control over the contribution and its underlying technology, in case of default.

12. Logsdon, *Together in Orbit,* 56.

13. NASA reported that it was "prudently developing contingency plans to allow the program to go forward in the event an international partner is unable to fulfill its obligations. Congressional representatives have endorsed the need for such planning in the case of Russia." Beth A. Masters, NASA director of International Relations, letter to U.S. Congress, Office of Technology Assessment, April 26, 1995, NASA Historical Reference Collection.

14. The issue of Russian reliability, NASA contingency plans, the reactions of foreign partners to Russia's inclusion in the program, and the general risks and benefits of U.S.-Russian space cooperation are discussed in Office of Technology Assessment, *U.S.-Russian Cooperation in Space.*

15. David M. Harland and John E. Catchpole, *Creating the International Space Station: Design, Assembly, and Utilization* (Chicester, England: Springer Verlag, 2002), chap. 12.

16. Matthew J. Von Bencke, *The Politics of Space: A History of U.S.-Soviet/Russian Competition and Cooperation* (Boulder, Col.: Westview Press, 1997), 107.

17. Lorenza Sebesta, "The Politics of Technological Cooperation in Space: U.S.-European Negotiations on the Post-Apollo Programme," *History and Technology: An International Journal* 11 (1994): 317–41.

18. William J. Clinton, "National Space Policy," September 19, 1996, NASA Historical Reference Collection; "Space Activities of the U.S. Government," in *Aeronautics and Space Report of the President* (Washington, D.C.: NASA, 2000), appendix E.

19. Roger D. Launius, "NASA, the Space Shuttle, and the Quest for Primacy in Space in an Era of Increasing International Competition," in Emmanuel Chadeau, ed., *"L'Ambition Technologique: Naissance d'Ariane* (Paris: Institut d'Histoire de l'Industrie, 1995), 35–61; Roger D. Launius, "The View from Washington," in E. A. Harris, ed., *The History of the European Space Agency: Proceedings of an International Symposium, 11–13 November 1998* (Noordwijk, the Netherlands: ESA Publications Division, SP-436, 1999), 201–15.

20. Kenneth S. Pedersen, "Thoughts on International Space Cooperation and Interests in the Post–Cold War World," *Space Policy* 8 (August 1992): 217.

21. Keith Cowing, ed., NASA Watch, written testimony for the Subcommittee on Space and Aeronautics, House Science Committee, "NASA at 40: What Kind of Space Program Does America Need for the Twenty-first Century?" October 1, 1998, available on-line at http://www.nasawatch.com/congress/10.01.98.cowing.html, accessed March 24, 2002.

22. Smith, "NASA's Space Station Program."

23. NASA Advisory Council, "Report of the Cost Assessment and Validation Task Force," 11.

24. NASA Johnson Space Center Fact Sheet, "The Zarya Control Module: The First International Space Station Component to Launch," January 1999; "Goldin Gives Russia Six Weeks to Get Station on Schedule," *Aerospace Daily,* March 27, 1996. Both in NASA Historical Reference Collection.

25. Piers Bizony, *Island in the Sky: Building the International Space Station* (London: Aurum Press, 1996), 95–97; personal communication with ISS official who asked for non-attribution, November 17, 2001.

26. NASA Johnson Space Center Fact Sheet, "A History of U.S. Space Stations," June 1997, NASA Historical Reference Collection.

27. NASA Advisory Council, "Report of the Cost Assessment and Validation Task Force."

28. Ibid.

29. NASA Johnson Space Center Fact Sheet, "The Zarya Control Module: The First International Space Station Component to Launch," January 1999, NASA Historical Reference Collection.

30. Ibid., 8.

31. NASA, "International Space Station User's Guide," Release 2.0, n.d., 10, NASA Historical Reference Collection; NASA Johnson Space Center Fact Sheet, "The International Space Station: An Overview," June 1999; NASA Johnson Space Center, STS-92, "Preflight Interview: Koichi Wakata," n.d. [October 2000], available on-line at http://spaceflight.nasa.gov/shuttle/archives/sts-92/crew/intwakata.html, accessed March 26, 2002; Computer Sciences Corporation, MOBIS Contract GS-23F-8029H, "International Space Station Operations Architecture Study, Final Report," August 2000, 1–2; NASA Ames Research Center Fact Sheet, ARC-97-04, "Gravitational Biology Facility and Centrifuge Facility on the International Space Station," September 1997, NASA Historical Reference Collection.

32. NASA Johnson Space Center Fact Sheet, "The International Space Station: An Overview."

33. NASA Johnson Space Center Fact Sheet, "The International Space Station: The First Steps to a New Home in Orbit," June 1999, NASA Historical Reference Collection.

34. NASA Johnson Space Center, "STS-96 Post-Mission Summary," available on-line at http://spaceflight.nasa.gov/shuttle/archives/sts-96/, accessed March 30, 2002.

35. NASA Johnson Space Center Fact Sheet, "The International Space Station: The First Steps to a New Home in Orbit"; "Zvezda: Cornerstone for Early Human Habitation of the International Space Station," NASA press kit, July 2, 2000, 30–33.

36. Smith, "NASA's Space Station Program"; Keith Cowing, "The Service Module Has Been Launched," SpaceRef.com, July 9, 2000, available on-line at http://www.spaceref.com/news/viewnews.html?id=177, accessed March 31, 2002.

37. NASA Johnson Space Center, "STS-101 Outfits International Space Station," available on-line at http://spaceflight.nasa.gov/shuttle/archives/sts-101/, accessed March 30, 2002. NASA Johnson Space Center, "100th Space Shuttle Flight," available on-line at http://spaceflight.nasa.gov/shuttle/archives/sts-92/, accessed March 30, 2002.

38. "Space News," *Space Times: Magazine of the American Astronautical Society* 40 (January–February 2001): 16; Launius, "2001: The Odyssey Continues," 3, 12.

39. NASA Johnson Space Center, "STS-97 Delivers Giant Solar Arrays to International Space Station," available on-line at http://spaceflight.nasa.gov/shuttle/archives/sts-97/, accessed March 30, 2002; NASA Johnson Space Center, "STS-98 Delivers Destiny Lab to International Space Station," available on-line at http://spaceflight.nasa.gov/shuttle/archives/sts-98/, accessed March 30, 2002; NASA Johnson Space Center Fact Sheet, "The International Space Station: The First Steps to a New Home in Orbit."

40. NASA, "International Space Station User's Guide," 9–10.

41. NASA Johnson Space Center, "STS-102 Swaps International Space Station Crews," available on-line at http://spaceflight.nasa.gov/shuttle/archives/sts-102/, accessed March 30, 2002; NASA Johnson Space Center, "STS-100 Delivers Canadarm2 to International Space Station," available on-line at http://spaceflight.nasa.gov/shuttle/archives/sts-100/, accessed March 30, 2002; NASA Johnson Space Center, "STS-104 Delivers Quest to International Space Station," available on-line at http://spaceflight.nasa.gov/shuttle/archives/sts-104/, accessed March 30, 2002.

42. NASA Johnson Space Center, "STS-105 Swaps International Space Station Crews," available on-line at http://spaceflight.nasa.gov/shuttle/archives/sts-105/, accessed March 30, 2002; NASA Johnson Space Center, "STS-108 Swaps International Space Station Crews," available on-line at http://spaceflight.nasa.gov/shuttle/archives/sts-108/, accessed March 30, 2002; "STS-110 Delivers Framework for Station Expansion," available on-line at http://spaceflight.nasa.gov/shuttle/archives/sts-110/index.html, accessed September 12, 2002.

43. NASA Advisory Council, "Report of the Cost Assessment and Validation Task Force," 11; Jim Shefter, "The Sum of Its Parts," *Popular Science* (May 1998): 3–7; Frank Oliveri, "NASA Mistakes, Optimism Cost Taxpayers," *Florida Today,* June 17, 2001; "Station's Cost More Than Triples Since Reagan Plan," *Florida Today,* June 17, 2001; "Space Station Costs Expected to Run $1 Billion a Year through 2017," *Florida Today,* September 6, 2002.

44. Marcia S. Smith, *Space Station,* CRS Issue Brief for Congress, IB93017, May 16, 2001, NASA Historical Reference Collection.

45. "The Sweet Smell of Air," *Sports Night,* first aired January 25, 2000.

46. R. Dale Reed with Darlene Lister, *Wingless Flight: The Lifting Body Story* (Washington, D.C.: NASA SP-4220, 1997),

187–91; NASA Dryden Flight Research Center, "X-38 CRV," February 13, 2001, available on-line at http://www.dfrc.nasa.gov/ Projects/X38/intro.html, accessed March 31, 2002; NASA Fact Sheet, "The X-38: Low-Cost, High-Tech Space Rescue," IS-2000-01-ISS022-JSC, 2000, NASA Historical Reference Collection; Eckart D. Graf, "ESA and the ISS Crew Return Vehicle," *On Station* (March 2001): 3–5.

47. On the lifting body program, see Reed with Lister, *Wingless Flight;* Milton O. Thompson and Curtis Peebles, *Flying without Wings: NASA Lifting Bodies and the Birth of the Space Shuttle* (Washington, D.C.: Smithsonian Institution Press, 1999).

48. NASA Dryden Flight Research Center Fact Sheet, "X-38," FS-2000-04-038 DFRC, April 2000, NASA Historical Reference Collection; Barton C. Hacker, "The Gemini Paraglider: A Failure of Scheduled Innovation, 1961–1964," *Social Studies of Science* 22 (spring 1992): 387–406. The paraglider was conceived during the 1950s as a lightweight hybrid of parachute and inflated wing that might allow astronauts to pilot spacecraft to airfield landings. From 1961 to 1964, NASA sought to convert the idea into a practical landing system for the Gemini spacecraft. The spacecraft would carry the paraglider safely tucked away through most of a mission. Only after reentering the atmosphere from orbit would the crew deploy the wing. Having converted the spacecraft into a makeshift glider, the crew could fly to an airfield landing. The system was later further developed, and by the 1990s it was compatible with spaceflight.

49. Hacker, "The Gemini Paraglider," 402–6; Reed with Lister, *Wingless Flight,* 131–43, 167–70; NASA Dryden Flight Research Center press release, "X-38 Crew Return Vehicle Prototype Resumes Flight Tests," June 28, 2001, NASA Historical Reference Collection.

50. NASA Dryden Flight Research Center Fact Sheet, "X-38," FS-2000-04-038 DFRC, April 2000; NASA Fact Sheet, "The X-38: Low-Cost, High-Tech Space Rescue"; Graf, "ESA and the ISS Crew Return Vehicle," 5.

51. NASA Dryden Flight Research Center Fact Sheet, "X-38," FS-2000-04-038 DFRC, April 2000; NASA Fact Sheet, "The X-38: Low-Cost, High-Tech Space Rescue"; Hacker, "The Gemini Paraglider," 387–406; NASA Johnson Space Center press release, "NASA X-38 Team Flies Largest Parafoil Parachute in History," February 3, 2000, NASA Historical Reference Collection; Graf, "ESA and the ISS Crew Return Vehicle," 5.

52. "Report by the International Space Station (ISS) Management and Cost Evaluation (IMCE) Task Force to the NASA Advisory Council," November 1, 2001, 7, NASA Historical Reference Collection.

53. Statement of Daniel S. Goldin, NASA administrator, before the Committee on Science House of Representatives, April 4, 2001, NASA Historical Reference Collection.

54. Frank Sietzen Jr. and Keith Cowing, "Habitation Use May Rescue Struggling Commercial Module Project," Space Ref.com, March 27, 2001, in NASA Historical Reference Collec-

tion; interview with Michael Hawes, March 16, 2002.

55. Smith, "NASA's Space Station Program."

56. "Report by the International Space Station (ISS) Management and Cost Evaluation (IMCE) Task Force to the NASA Advisory Council," 5; NASA Advisory Council, "Report of the Cost Assessment and Validation Task Force," 2.

57. Jefferson Morris, "NASA Administrator Expresses Support for ISS Task Force Recommendations," *Aerospace Daily,* November 7, 2001; Brian Berger, "Canadians, Europeans Warn U.S. on Space Station," November 7, 2001, Space.com, all in NASA Historical Reference Collection; "A Space Station Out of Control," *New York Times,* November 25, 2001.

58. "House Science Committee Hearing Charter: The Space Station Task Force Report," Committee on Science, U.S. House of Representatives, November 7, 2001, NASA Historical Reference Collection; interview with senior NASA ISS project member who asked for non-attribution, November 21, 2001.

59. "A Space Station Out of Control," *New York Times.*

60. "House Science Committee Hearing Charter: The Space Station Task Force Report."

61. Ibid.; "Report by the International Space Station (ISS) Management and Cost Evaluation (IMCE) Task Force to the NASA Advisory Council."

62. Keith Cowing, "NASA's Budget: Back to the Future— and to Basics," February 5, 2002, SpaceRef.com, NASA Historical Reference Collection.

63. "Report by the International Space Station (ISS) Management and Cost Evaluation (IMCE) Task Force to the NASA Advisory Council," 4.

64. Ibid., 5.

65. Aaron Cohen, "Project Management: JSC's Heritage and Challenge," *Issues in NASA Program and Project Management* (Washington, D.C.: NASA SP-6101, 1989), 7–16; C. Thomas Newman, "Controlling Resources in the Apollo Program," *Issues in NASA Program and Project Management* (Washington, D.C.: NASA SP-6101, 1989), 23–26; Eberhard Rees, "Project and Systems Management in the Apollo Program," *Issues in NASA Program and Project Management* (Washington, D.C.: NASA SP-6101 (02), 1989), 24–34.

66. NASA, "Key Space Station Messages," July 10, 2001, copy in possession of author.

7. EPILOGUE: WHITHER THE INTERNATIONAL SPACE STATION

1. NASA Johnson Space Center Fact Sheet, "International Space Station Assembly: A Construction Site in Orbit," IS-1999-06-ISS013JSC, June 1999, NASA Historical Reference Collection.

2. Interview with Robert C. Treviño, November 15, 2000; Portree and Treviño, *Walking to Olympus: A Chronology of Extravehicular Activity (EVA).*

3. "STS-104 Delivers Quest to International Space Station," http://spaceflight.nasa.gov/shuttle/ archives/sts-104/index.html; Lillian D. Kozloski, *U.S. Space Gear: Outfitting the Astronauts* (Washington, D.C.: Smithsonian Institution Press, 1993), 129–44.

4. NASA Johnson Space Center Fact Sheet, "International Space Station Assembly: A Construction Site in Orbit."

5. Arnauld E. Nicogossian, Deborah F. Pober, and Stephanie A. Roy, "Evolution of Telemedicine in the Space Program and Earth Applications," *Telemedicine Journal and E-Health* 7 (2001): 1–15.

6. "A Space Station Out of Control," *New York Times.*

7. Interview with senior NASA science official who asked for non-attribution, July 16, 2002.

8. "A Space Station Out of Control," *New York Times.*

9. Hon. Ralph M. Hall, speech to the AAS Goddard Memorial Symposium, March 27, 2001, Greenbelt, Md., published as "'A Time of Transition': Remarks to the American Astronautical Society's Goddard Memorial Symposium," *Space Times: Magazine of the American Astronautical Society* 40 (September–October 2001): 12–15.

10. A. Thomas Young, "Report by the International Space Station (ISS) Management and Cost Evaluation (IMCE) Task Force to the NASA Advisory Council," November 1, 2001, 29, NASA Historical Reference Collection.

11. NASA Office of Biological and Physical Research, *OBPR Near Term ISS Research Priority Definitions, 2002–2006* (Washington, D.C: NASA, June 2001), 1.

12. Arnauld Nicogossian and Deborah Pober, "The Future of Space Medicine," *Acta Astronautica* 49 (2001): 529–35; Arnauld E. Nicogossian, "Human Factors for Mars Missions," in D. B. Reiber, ed., *The NASA Mars Conference* (San Diego, Calif.: Univelt, AAS Science and Technology Series, vol. 71, 1988), 475–86.

13. NASA Office of Biological and Physical Research, "OBPR Near Term ISS Research Priority Definitions, 2002–2006," June 2001; NASA Office of Biological and Physical Research, "ISS Research Plan: Research Increments 1-5," 2001; NASA Office of Biological and Physical Research, "Science and Technology Research Directions for the International Space Station," January 2000. All in NASA Historical Reference Collection.

14. NASA Office of Biological and Physical Research, "Research on the International Space Station," June 2001, NASA Historical Reference Collection."

15. "NASA Administrator to Make Space Station Research Announcement," September 13, 2002, NASA Historical Reference Collection.

16. NASA Office of Biological and Physical Research, "Research on the International Space Station," June 2001, NASA Historical Reference Collection."

17. "Space Station Science," available on-line at http://spaceflight.nasa.gov/station/science/experiments/index.html, accessed April 8, 2002.

18. NASA Office of Biological and Physical Research, "International Space Station: Payload Operations Concepts and Architecture Assessment Study," February 2002, NASA Historical Reference Collection.

19. Quoted in "On Bureaucracy," *Chicago Sun Times,* July 10, 1958; quoted in Arnold S. Levine, *Managing NASA in the Apollo Era* (Washington, D.C.: NASA SP-4102, 1982), v.

20. See Roger D. Launius, *Apollo: A Retrospective Analysis* (Washington, D.C.: Monographs in Aerospace History, no. 3, 1994).

21. On the reorientation of world politics in the 1990s, see John Lewis Gaddis, "Toward the Post–Cold War World," *Foreign Affairs* 70 (spring 1991): 101–14; Judith Goldstein and Robert O. Keohane, ed., *Ideas and Foreign Policy: Beliefs, Institutions, and Political Change* (Ithaca, N.Y.: Cornell University Press, 1993); Francis Fucayama, "The End of History," *The National Interest* 16 (summer 1989): 3–18; Max Singer and Aaron Wildavsky, *The Real World Order: Zones of Peace, Zones of Turmoil* (Chatham, N.J.: Chatham House, 1993); James M. Goldgeier and Michael McFaul, "A Tale of Two Worlds: Core and Periphery in the Post–Cold War Era," *International Organization* 46 (spring 1992): 467–91; Kenneth N. Waltz, "The Emerging Structure of International Politics," *International Security* 18 (fall 1993): 44–79; Zbigniew Brzezinski, *Out of Control: Global Turmoil on the Eave of the Twenty-first Century* (New York: Scribner, 1993); Daniel Patrick Moynihan, *Pandaemonium: Ethnicity in International Politics* (New York: Oxford University Press, 1993); William S. Lind, "North-South Relations: Returning to a World of Cultures in Conflict," *Current World Leaders* 35 (December 1993): 1073–80; Donald J. Puchala, "The History of the Future of International Relations," *Ethics and International Affairs* 8 (1994): 177–202.

22. This provocative thesis is illuminated in Samuel P. Huntington, *The Clash of Civilizations and the Remaking of World Order* (New York: Simon and Schuster, 1997).

23. Huntington, *The Clash of Civilizations,* 40–78.

24. Ibid., 266–98; "China Publishes Plans for Space Exploration," *Voice of America News,* November 22, 2000; Marc Boucher, "Shenzhou 2 Launch Imminent, Chinese Manned Space Program Targets the Moon," *New York Times,* October 30, 2000; People's Republic of China, Information Office of the State Council, "China's Space Activities," November 22, 2000, in NASA Historical Reference Collection.

25. Gerard K. O'Neill, "The Colonization of Space," *Physics Today* 27 (September 1974): 32–40; Gerard K. O'Neill, *The High Frontier: Human Colonies in Space* (New York: William Morrow, 1976); Peter E. Glaser, "Energy from the Sun: Its Future," *Science* 162 (1968): 857–60; Peter E. Glaser, "Solar Power via Satellite," *Astronautics and Aeronautics* (August 1973): 60–68; Peter E. Glaser, "An Orbiting Solar Power Station," *Sky and Telescope* (April 1975): 224–28.

26. Arthur C. Clarke, *Rendezvous with Rama* (New York: Bantam Books, 1973); T. A. Heppenheimer, *Colonies in Space*

(Harrisburg, Pa.: Stackpole Books, 1977).

27. This would be completely consistent with their ideology. See Roger D. Launius, "A Western Mormon in Washington, D.C.: James C. Fletcher, NASA, and the Final Frontier," *Pacific Historical Review* 64 (May 1995): 217–41; Hans Mark, *The Space Station: A Personal Journey* (Durham, N.C.: Duke University Press, 1987); "Colonies in Space," *Newsweek* (November 27, 1978): 95–101.

28. See Richard D. Johnson and Charles Holbrow, eds., *Space Settlements: A Design Study in Colonization* (Washington, D.C.: NASA SP-413, 1977).

29. The latter half of the 1970s might best be viewed as a nadir in human space exploration, with the Apollo program gone and the Shuttle not yet flying. See Louis J. Halle, "A Hopeful Future for Mankind," *Foreign Affairs* 59 (summer 1980): 1129–36.

30. Freeman Dyson, "Obituary, Gerard Kitchen O'Neill," *Physics Today* 46 (February 1993): 97–98.

31. John M. Logsdon, "The Space Station is Finally Real," *Space Times: Magazine of the American Astronautical Society* 34 (November–December 1995): 23.

32. Ibid.

33. For a fuller discussion of this subject, see Launius and McCurdy, *Imagining Space,* 94–98.

34. Clarke, *2001: A Space Odyssey.*

35. Jim Banke, "Russian Soyuz Rocket Lifts Progress Toward Mir, Space.com, October 16, 2000, available on-line at http://space.com/missionlaunches/launches/soyuz_progress_launch_001016.html, accessed August 14, 2002.

36. Patrick Collins, "The Space Tourism Industry in 2030," in Stewart W. Johnson, Koon Meng Chua, Rodney G. Galloway, and Philip J. Richter, eds., *Space 2000: Proceedings of the Seventh International Conference and Exposition on Engineering, Construction, Operations, and Business in Space* (Reston, Va.: American Society of Civil Engineers, 2000), 594–603; interview with Roy W. Estess, former director, Johnson Space Center, June 25, 2002; Dwayne A. Day, "From Astropower to Everyman to Rich Man: The Changing Human Face of Spaceflight," *Space Times: Magazine of the American Astronautical Society* 40 (July–August 2001): 22–23.

37. Estess interview, June 25, 2002.

38. Albert A. Harrison, "Our Future beyond Earth," *Space Times: Magazine of the American Astronautical Society* 40 (July–August 2001): 12.

39. "SPACE.com Survey Reveals Strong Public Support for Dennis Tito's Flight," Space.com, May 7, 2001, available on-line at http://www.space.com/news/tito_poll_010507.html, accessed August 14, 2002.

40. Ibid.; Day, "From Astropower to Everyman to Rich Man," 22–23.

41. This issue of space access is critical to opening any part of space to broad usage. See Roger D. Launius and Lori B. Garver, "Between a Rocket and a Hard Place: Episodes in the Evolution of Launch Vehicle Technology," IAA-00-IAA.2.2.02, paper presented at the 51st International Astronautical Congress, October 2–6, 2000, Rio de Janeiro; Roger D. Launius and Dennis R. Jenkins, eds., *To Reach the High Frontier: A History of U.S. Launch Vehicles* (Lexington: University of Kentucky Press, 2002); Howard E. McCurdy, "The Cost of Space Flight," *Space Policy* 10 (November 1994): 277–89; Craig R. Reed, "Factors Affecting U.S. Commercial Space Launch Industry Competitiveness," *Business and Economic History* 27 (fall 1998): 222–36.

BIBLIOGRAPHICAL ESSAY

A book such as this benefits from the previous efforts of many. There are numerous general studies of spaceflight that are important in understanding the development of the field in the twentieth century. By far the most significant is Walter A. McDougall, *The Heavens and the Earth: A Political History of the Space Age* (New York: Basic Books, 1985. Reprint, Baltimore: Johns Hopkins University Press, 1997), which received the Pulitzer Prize for its analysis of the space race to the Moon in the 1960s. The author argues that Apollo prompted the space program to stress engineering over science, competition over cooperation, civilian over military management, and international prestige over practical applications. Nearly as important is William E. Burrows, *This New Ocean: The Story of the First Space Age* (New York: Random House, 1998), which presents a strong overview of the history of the space age from Sputnik to 1998. Additionally, T. A. Heppenheimer's *Countdown: The History of Space Exploration* (New York: John Wiley and Sons, 1997) presents a general history, which is somewhat quirky but well written and entertaining. Although encompassing subjects broader than spaceflight, Roger E. Bilstein's *Flight in America: From the Wrights to the Astronauts* (Baltimore: Johns Hopkins University Press, 1984, 3d rev. ed. 2001) offers a superb synthesis of the origins and development of aerospace activities in the United States. These are the books to start with in any investigation of air and space activities.

There are five general histories of space stations that are of merit, but all are unsatisfactory in their efforts to explain the evolution of the concept. The best is Giovanni Caprara, *Living in Space: From Science Fiction to the International Space Station* (Willowdale, Ontario: Firefly Books, 2000). It tells the story of the orbital pioneers of the past and envisions a mythical fascinating future. Narrating this century-old story, the book combines science fiction, scientific and technological breakthroughs, political intrigue, and imaginative ideas. A second work is Piers Bizony, *Islands in the Sky: Building the International Space Station* (London: Aurum Press, 1996), a solid popular account of the possibilities of space stations, from their beginnings to the present effort. In spite of its title, which suggests a future-tense focus, *Challenges of Human Space Exploration* (Chicester, England: Springer-Praxis, 2000), by Marsha Freeman, deals largely with the series of space station efforts of the United States and the Soviet Union/Russia from the 1970s to the recent past. G. Harry Stine, *Living in Space: A Handbook for Work and Exploration Stations* beyond the Earth's Atmosphere (New York: M. Evans, 1997), explains the technology necessary for staying alive and the basic problems of working in space. Finally, several essays in John Zukowsky, ed., 2001: Building for Space Travel (New York: Harry N. Abrams, 2001), present an interesting discussion of the aesthetics of space stations. Howard E. McCurdy's study Space and the American Imagination (Washington, D.C.: Smithsonian Institution Press, 1997) is an enormously significant analysis of the relationship between popular culture and public policy, including the role of space stations.

On the early dreams of the spaceflight pioneers and the need for a space station, the place to start is Hermann Noordung, The Problem of Space Travel: The Rocket Motor, Ernst Stuhlinger and J. D. Hunley, eds., with Jennifer Garland (Washington, D.C.: NASA SP-4026, 1995), which provides the earliest technical description of a station. The contributions of Konstantin Tsiolkovskiy and others are evident in Konstantin E. Tsiolkovskiy, Works on Rocket Technology (Washington, D.C.: NASA, TT F-243, 1965); Arkady Kosmodemyansky, Konstantin Tsiolkovskiy (Moscow: Nauka, 1985); and Peter A. Gorin, "Rising from a Cradle: Soviet Public Perceptions of Space Flight before Sputnik," in Roger D. Launius, John M. Logsdon, and Robert W. Smith, eds., Reconsidering Sputnik: Forty Years Since the Soviet Satellite (Amsterdam: Harwood Academic Publishers, 2000). On the place of Hermann Oberth in this era, see his Rockets into Planetary Space (Washington, D.C.: NASA, TT F-9227, 1972), and H. B. Walters, Hermann Oberth: Father of Space Travel (New York: Macmillan, 1962). General discussions of the early era may be found in Frank H. Winter, Prelude to the Space Age: The Rocket Societies, 1924–1940 (Washington, D.C.: Smithsonian Institution Press, 1983); Roger D. Launius, Frontiers of Space Exploration (Westport, Conn.: Greenwood Press, 1998); Roger D. Launius, "Prelude to the Space Age," in John M. Logsdon, gen. ed., Exploring the Unknown: Selected Documents in the History of the U.S. Civil Space Program, vol. 1, Organizing for Exploration (Washington, D.C.: NASA SP-4407, 1995); and Arthur C. Clarke, ed., The Coming of the Space Age (New York: Meredith Press, 1967).

There is a wealth of literature on the development of spaceflight in the 1950s and 1960s and on the central role played by Werhner von Braun and his integrated plan for exploring space. As an entrée to this subject, see Frederick I. Ordway III and Randy Lieberman, eds., Blueprint for Space: Science Fiction to Science Fact (Washington, D.C.: Smithsonian Institution Press, 1992). Cornelius Ryan, ed., Across the Space Frontier (New York: Viking Press, 1952), publishes the best work on spaceflight from Collier's magazine, including Wernher von Braun's concept for a space station. The standard work on Sputnik is Rip Bulkeley, The Sputniks Crisis and Early United States Space Policy: A Critique of the Historiography of Space (Bloomington: Indiana University Press, 1991), an important discussion of early efforts to develop space policy in the aftermath of the Sputnik crisis of 1957.

A classic study of the critical decision in space policy is John M. Logsdon, The Decision to Go to the Moon: Project Apollo and the National Interest (Cambridge: MIT Press, 1970), an analysis of the political process in the United States leading to the decision to go to the Moon in 1961. This decision derailed the space station as a possibility in NASA's plans for more than a decade. The standard work on the exploration of the Moon is Andrew Chaikin, A Man on the Moon: The Voyages of the Apollo Astronauts (New York: Viking Press, 1994), which emphasizes the actions of the astronauts between 1968 and 1972.

There have been several useful works on Skylab, the American effort to build an orbital workshop as a precursor to a space station in the 1970s. W. David Compton and Charles D. Benson, Living and Working in Space: A History of Skylab (Washington, D.C.: NASA SP-4208, 1983), is the official NASA history of the project. Henry S. F. Cooper, A House in Space (New York: Holt, Rinehart, and Winston, 1976), presents a journalist's "I was there" type of account about Skylab. David J. Shayler's Skylab: America's Space Station (Chicester, England: Springer-Praxis, 2001), provides an overview. The history of the Soviet Salyut space stations has been discussed in Phillip Clark, The Soviet Manned Space Programme: An Illustrated History of the Men, the Missions and the Spacecraft (London: Salamander Books, 1988); Brian Harvey, The New Russian Space Programme: From Competition to Collaboration (Chicester, England: Springer-Praxis, 1996); Matthew J. Von Bencke, The Politics of Space: A History of U.S.-Soviet/Russian Competition and Cooperation in Space (Boulder, Col.: Westview Press, 1997); and most importantly in the masterful work by Asif A. Siddiqi, Challenge to Apollo: The Soviet Union and the Space Race, 1945–1974 (Washington, D.C.: NASA SP-2000-4408, 2000).

The U.S. effort in the 1980s to build a space station has been related in several important studies. Howard E. McCurdy prepared a fine study of the political process that led to the presidential decision in 1984 to develop an orbital station, The Space Station Decision: Incremental Politics and Technological Choice (Baltimore: Johns Hopkins University Press, 1990). An overview of the international cooperation involved in this project is presented in John M. Logsdon, *Together in Orbit: The Origins of International Cooperation in the Space Station Program* (Washington, D.C.: Monographs in Aerospace History, no. 11, 1998). A European perspective may be found in Roger M. Bonnet and Vittorio Manno, *International Cooperation in Space: The Example of the European Space Agency* (Cambridge: Harvard University Press, 1994). An excellent unpublished study of the Space Station Freedom program is Adam L. Gruen, "The Port Unknown: A History of Space Station Freedom" (NASA Historical Reference Collection, NASA History Office, Washington, D.C., 1992). Anyone interested in the history of Freedom would benefit from using this work as a starting point. A study of the history of Freedom's management is available in Thomas J. Lewin and V. K. Narayanan, *Keeping the Dream Alive: Managing the Space Station Program, 1982–1986* (NASA Contractor Report 4272, 1990), also available in the NASA Historical Reference Collection.

There are several works on the Mir space station operated between 1986 and 2001. David M. Harland's study, *The Mir Space Station: A Precursor to Space Colonization* (New York: John Wiley and Sons, 1997), is a solid descriptive history of the Soviet/Russian space station and the activities aboard it. See also Neal Bernards, *Mir Space Station* (New York: Smart Apple Media, 1999). Specialized discussions of parts of the Mir story are told in Bart Hendrickx, "The Origins and Evolution of Mir and its Modules," *Journal of the British Interplanetary Society* 51 (June 1998): 203–22; David J. Shayler, "Outside Mir: Ten Years of EVA Operations," *Journal of the British Interplanetary Society* 51 (January 1998): 29–38; and Vladimir Pivynuk and Mark Bock, *Space Station Handbook: Mir User's Manual* (Washington, D.C.: Cosmos Books, 1994).

The Shuttle-Mir program has received considerable coverage. An illustrated history, containing a CD/ROM with oral histories, documents, and multimedia materials, is Clay Morgan, *Shuttle-Mir: The U.S. and Russia Share History's Highest Stage* (Washington, D.C.: NASA SP-2001-4225, 2001). Bryan Burrough's *Dragonfly: NASA and the Crisis Aboard the Mir* (New York: Ballinger, 1998) provides a journalistic analysis of the American-Russian cooperation in space in the mid-1990s aboard the Mir space station. It emphasizes the crises on Mir that took place in the spring and summer of 1997, especially the fire and the collision of Progress with Spektr. Burrough culls his story from one-on-one interviews and transcripts of recorded conversations between the astronauts and cosmonauts on Mir and Russian Mission Control. He delves deeply into the personal and professional lives of the eleven people who lived aboard Mir from 1995 to 1998 and writes a simultaneously disheartening and fascinating story.

Two memoirs are Jerry M. Linenger, *Off the Planet: Surviving Five Perilous Months aboard the Space Station Mir* (New York: McGraw-Hill, 1999), the story of an astronaut who lived on Mir during the crises of 1997; and Colin Foale, *Waystation to the Stars: The Story of Mir, Michael, and Me* (London: Headline, 1999), an account by astronaut Michael Foale's father.

The possibilities for space colonies have been discussed in several works. Gerard K. O'Neill started the craze in 1972 with the publication of "The Colonization of Space," *Physics Today* 27 (September 1974): 32–40. He followed that with a highly influential book, *The High Frontier: Human Colonies in Space* (New York: William Morrow, 1976), which energized a political interest group. T. A. Heppenheimer's *Colonies in Space* (Harrisburg, Pa.: Stackpole Books, 1977) expanded on O'Neill's ideas. See also Hans Mark, *The Space Station: A Personal Journey* (Durham, N.C.: Duke University Press, 1987); and Stanley Schmidt and Robert Zubrin, *Islands in the Sky: Bold New Ideas for Colonizing Space* (New York: John Wiley and Sons, 1996). Richard D. Johnson and Charles Holbrow, eds., *Space Settlements: A Design Study in Colonization* (Washington, D.C.: NASA SP-413, 1977), presents the findings of a study sponsored by NASA Ames, American Society of Electrical Engineers, and Stanford University in the summer of 1975 to look at all aspects of sustained life in space. See also John Billingham, William Gilbreath, Gerard K. O'Neill, and Brian O'Leary, eds., *Space Resources and Space Settlements* (Washington, D.C.: NASA SP-428, 1979).

A discussion of possible futures in space may be found in Roger D. Launius and Howard E. McCurdy, *Imagining Space: Achievements, Predictions, Possibilities, 1950–2050* (San Francisco: Chronicle Books, 2001). Frank White, *The Overview Effect: Space Exploration and Human Evolution,* 2d ed. (Reston, Va.: American Institute of Aeronautics and Astronautics, 1998), provides a significant analysis and exposition of why humanity must explore space. Carl Sagan, *Pale Blue Dot: A Vision of the Human Future in Space* (New York: Random House, 1994), presents probably the most sophisticated articulation of the space exploration imperative since Wernher von Braun's work of the 1950s and 1960s.

Page numbers in boldface indicate illustrations.

262